ROUTLEDGE LIBRARY EDITIONS:
URBAN PLANNING

Volume 3

LAND RENT, HOUSING AND URBAN PLANNING

LAND RENT, HOUSING AND URBAN PLANNING

A European Perspective

Edited by
M. BALL, V. BENTIVEGNA,
M. EDWARDS AND M. FOLIN

LONDON AND NEW YORK

First published in 1985 by Croom Helm Ltd

This edition first published in 2018
by Routledge
2 Park Square, Milton Park, Abingdon, Oxon OX14 4RN

and by Routledge
711 Third Avenue, New York, NY 10017

Routledge is an imprint of the Taylor & Francis Group, an informa business

British Library Cataloguing in Publication Data
A catalogue record for this book is available from the British Library

ISBN: 978-1-138-49611-8 (Set)
ISBN: 978-1-351-02214-9 (Set) (ebk)
ISBN: 978-1-138-49443-5 (Volume 3) (hbk)
ISBN: 978-1-138-49447-3 (Volume 3) (pbk)
ISBN: 978-1-351-02614-7 (Volume 3) (ebk)

Correction to Table 2.1, Chapter 2

The following table is a correction to Table 2.1, which appears on page 31 of chapter 2 of the publication. The authors have kindly redrawn it for this 2018 reissue.

Table 2.1: Nomenclature and Definitions for the Analysis of the Composition of New House Prices – Corrected version

Individual categories	Aggregated categories						
1 Site price paid to seller	13 Site acquisition cost		15 Cost of developed site				
2 Site transaction costs (survey, tax, legal, disturbance, demolitions, evictions							
3 Spending on network services, roads		14 Site development expenditure		17 Technical development expenditure			
4 Contributions to public services (local charges for infrastructure, cost of 'planning gain", free land transfers etc)							
5 Site preparation costs (leveling, landscaping, garages and parking spaces							
6 Building expenditure incl foundations			16 Construction cost		19 Total development expenses		
7 Fees relating to building (architect, QS, planning)						20 Selling Price	
8 Financing costs				18 Overhead expenses			21 Gross development surplus
9 Any residual VAT							
10 Developer's management							
11 Marketing							
12 Net development surplus (or apparent development surplus: return on development capital, pre-tax)							

Note: The terms 'cost', 'price', 'expenditure' are used more or less interchangeably here (as in the French). The conceptual problems of terminology are discussed at length in the text. The original version of this table on page 31 also has the French terms, but lacked the tones and boxes which indicate the composition of the aggregates. (Eds, 2018).

Land Rent, Housing and Urban Planning

A EUROPEAN PERSPECTIVE

Edited by
M. BALL, V. BENTIVEGNA,
M. EDWARDS and M. FOLIN

CROOM HELM
London • Sydney • Dover, New Hampshire

Croom Helm Ltd, Provident House, Burrell Row,
Beckenham, Kent BR3 1AT
Croom Helm Australia Pty Ltd, First Floor, 139 King Street,
Sydney, NSW 2001, Australia

British Library Cataloguing in Publication Data

Land, rent, housing and urban planning.
1. Marx, Karl, *1818-1883* 2. Rent
3. Rent (Economic theory)
I. Ball, M.
333.33'8 HD7287.5

ISBN 0-7099-3240-5

Croom Helm, 51 Washington Street, Dover,
New Hampshire, 03820, USA

Library of Congress Cataloging in Publication Data
Main entry under title:

Land,rent, housing, and urban planning.

Bibliography: p.
Includes index.
1. Rent (Economic theory) – Congresses. 2. Leases – Europe – Congresses. 3. Land use, Urban – Europe – Congresses. 4. Marxian economies – Congresses. 5. Housing policy – Europe – Congresses. 6. Urban policy – Europe – Congresses. I. Ball, Michael.
HB401.L36 1984 333.5'094 84-17497
ISBN 0-7099-3240-5

Typeset by Mayhew Typesetting, Bristol, England
Printed and bound in Great Britain by
Biddles Ltd, Guildford and King's Lynn

CONTENTS

TABLES

FIGURES

ACKNOWLEDGEMENTS

This volume comprises the proceedings of the symposium on land ownership held in Florence in December 1982. The symposium and the publication of this book have been supported financially by the Istituto di Urbanistica of the Faculty of Architecture, University of Florence. The symposium enabled all the contributors to discuss each other's work and thus to complete their sections in the light of collective, as well as editorial, comments. Thanks are also due to the the Bartlett International Summer Schools on the Production of the Built Environment for support and for providing the context in which some of the themes of the book have been developed.

Michael Ball
Vincenzo Bentivegna
Michael Edwards
Marino Folin

PART ONE:
EDITORS' INTRODUCTION

1 MODERN CAPITALISM AND THE THEORY OF URBAN RENT: A REVIEW

Introduction

Theories of land rent focus traditionally on agriculture. The Marxist theory of rent is no exception. Yet over the past fifteen years a broad literature within the Marxist tradition has emerged which looks at the non-agricultural contexts in which land rent is important. One strand in this literature is the examination of the impact of rent on particular industries – like those involved in minerals extraction and the construction industry – the physical content of whose production processes brings them in to an immediate and uncomfortable relation to landed property. The vast urban agglomerations of modern capitalist societies raise similar issues. In them all activities have to come to terms with the fact that landed property can extract high rents from urban areas. So the question arises of the extent to which land rent contributes to the enormous problems that beset urban life. Attempts to deal with particular urban problems via state intervention, over housing provision for instance, or infrastructure expenditure and land use planning, have to overcome the opposition of landed property and frequently they fail to do so.

This book presents a series of essays which try to develop our understanding of the role of landed property in the non-agricultural context. They are representative of the broad trends of the most recent theoretical and empirical developments. A wide range is covered ranging historically from Ancient Rome; through the 'golden years' of *laissez-faire* capitalism in essays dealing with the nineteenth-century British coal industry and disputes over the legal definition of landed property during the formative years of the British planning system in the late nineteenth and early twentieth centuries; the debate on Marx's categories of rent then disrupts the linear trend of history extending as it does from the time Marx wrote to the present day; but consideration of capitalist urbanisation and landed property takes up the historical theme again with essays dealing with the broad trends of urban development through to contemporary debates over political strategies towards urban land use and ownership. The juxtaposition of such a wide range of issues on rent has been deliberate as each area throws up new understandings of the role of rent which then are mutually reinforcing in terms of the theoretical insights they provide, and of the limitations that they reveal in certain perspectives.

The contributors also come from three different European countries, France, Great Britain and Italy, which adds to the breadth of the perspective. Not only do these authors come from countries with different economic, social and political backgrounds, but these three European countries are also the ones in which there has been the greatest debate over land rent questions during the past fifteen years.

By far the greatest interest in Marx's theory of rent in recent years has been associated with the urban context, and the balance of articles in this book reflects this trend. For this reason we shall undertake two interrelated tasks in this introductory chapter. The first is to survey the developments in the use of Marx's theory of rent in the urban context, and the second is to provide a brief synopsis of the chapters that follow.

Marxist Rent Theory and Urban Problems[1]

The Marxist theory of rent contrasts with the rent theories of the classical economists, such as Smith and Ricardo, and with modern neoclassical economics in rejecting the idea that land rent has a neutral, residual role in capitalist societies. Marxist rent theory emphasises instead the social nature of the rent relation and the importance of historical change. Rent is a revenue appropriated by landowners out of the surplus value capitalists exploit from the working class in the process of capital accumulation. The rent relation is consequently structured by the dynamic of the accumulation process. Variations in the rate and form of accumulation and the concrete social circumstances in which accumulation is occurring, all in various ways structure the effects of the rent relation. Landed property, moreover, might at times aid the process of accumulation, most spectacularly in the genesis of industrial capitalism in eighteenth-century Britain, whereas at other times it might hinder accumulation. Modern urban life seems to provide a fertile context in which rent can act as such a barrier, directly via its restrictions on the profitability of investment and indirectly via the divisive social cleavages it can open up. The focus of attention in recent years on urban rent relations is not, therefore, surprising.

Marx wrote extensively on the theory of rent in *Capital* and *The Theories of Surplus Value*. His writings on the subject are inevitably open to different interpretations, and a considerable literature has subsequently emerged on what the correct interpretation should be. Marx, however, had little to say about urban rent. Considerable interest was generated in Marx's theory of rent at the turn of the century as revolutionary movements tried to grapple with the problems of agrarian societies and of the means by which to forge alliances between peasants and the working class. Lenin and Kautsky provided the major interventions into this debate, but it did not spark off a comparable debate over the urban situation. In a number of respects, this lack of analysis of urban problems was a serious omission from revolutionary strategies. It reflected an over-optimism about the ease with which a politically-united revolutionary working class could be forged, and by examining the point of production in isolation neglected the consequences of the acute urban crisis which was affecting the daily lives of broad sections of the population. It was comparatively easy, as reformist social democratic tendencies were to discover, to link popular concern over living conditions to critiques of landed property and, hence, of private property as a whole. Though in the case of such political tendencies, the critique of private property was half-hearted to say the least. Yet, for revolutionary socialism, an opportunity was lost to use the problems of urban life to create a community of interest which broke through the status and gender divisions rending the working class.

Little further development in the analysis of rent occurred during the 50 years after 1917. But the late 1960s and early 1970s saw a considerable revival of interest in Marxist theory and a determination amongst many to learn from the lessons of past mistakes, even if that necessitated substantial criticisms of accepted political wisdoms and the theories which underlay them. A revival of interest in the theory of rent was one of the outcomes of that project. Three separate areas stimulated enquiry. The position of Third World countries in the world capitalist system helped to generate two of them. The role of land rent in the creation of underdevelopment and dependency was examined, albeit with only partial success. The position of the Third World as a provider of raw materials, in combination with the growing energy crisis in advanced capitalist countries, encouraged interest in the application of Marx's rent theory to minerals extraction. The third area was associated with urban problems. The political situation in advanced capitalist countries at the time was a major stimulus to investigations in this area.

The late 1960s and early 1970s were therefore years which brought together a series of influences that sparked off considerable interest in theories of urban rent. There was a manifest crisis in many aspects of daily urban life. The years of the long post-war boom had brought considerable prosperity, but they had also left substantial pockets of poverty concentrated particularly in inner urban areas, and led to little or no tangible improvement in the urban living conditions of most working people. 'Full' employment had been achieved (even if it was to be shortlived) and a welfare state of sorts had emerged in most advanced capitalist societies. Yet the forms of prosperity that had grown up during the 1950s and 1960s created widespread dissatisfaction. Status and gender divisions had not been broken down; if anything they had been reinforced. The bureaucratic forms associated with welfare/Keynesian state ideologies led to disillusionment with the consensus political strategies of the post-war era. Social control rather than social advancement became more clearly understood as their guiding principle. The new forms of mass housing provision, the massive expansion of car ownership and the road systems that went with it, the wholesale destruction of large city areas in the name of rational planning, the clear attempts to use the local apparatus of the state to promote racist ideologies and spatial segregation, and the frequent absence of essential social facilities sparked off considerable discontent. Spontaneous protest movements emerged in response to these failings in the cities of most advanced capitalist countries. Usually their activities were sporadic, single-issue and unorganised, but a new social category had come into sharp political prominence: protest movements that came to be known in the literature as urban social movements.

The closing years of the long post-war boom brought to a head another growing tendency of the 1950s and 1960s which could be counterposed to the problems that many faced in their daily lives: land and property speculation. The restructuring of the capitalism system that had taken place throughout the post-war years produced a number of tendencies which brought property speculation into prominence. The growth in living standards had gone hand in hand with substantial suburbanisation, although the forms it took varied considerably from country to country. Agricultural land at the fringes of urban areas became ripe for speculation and urban development by a variety of land and property interests. Whereas, in the inner city areas, the limitations of surburban lifestyles encourage the displacement of poorer households by richer ones in a process that has come to be known as gentrification. Cleavages and tensions over housing provision could consequently be clearly linked to property interests out to make a profit. Rent analysis seemed a fruitful way of exploring those contradictions and struggles.

Housing provision was not the only sphere where land and property speculation gained political prominence. The growing centralisation and complexity of capitalist enterprises, the increasingly international character of capitalist production and the mounting sophistication and importance of the world financial system led to an enormous demand for prestigious central city offices. Substantial transformations of central city areas to meet these needs were undertaken from the 1950s onwards. Such transformations contributed to the fragmented and truncated nature of urban life, and in combination with suburbanisation, generated long journeys to work and a spatially separated and divided urban populace. Yet the end of the long post-war boom produced a further twist to the growth of post-war property markets in the contradictions that it threw up between the financial sector and other spheres of accumulation. The financial sector in the post-war years retained its earlier pre-1945 links with governments and industrial capital, but the steady expansion of international trade and commodity speculation, of property markets, and of personal savings linked to insurance and pension schemes, created substantial new dimensions. The decline of industrial profitability and investment at the end of the post-war boom in the late 1960s/early 1970s coincided with, and partially caused, an enormous expansion in credit. Inflationary pressures mounted with considerable consequences for housing finance and for land and property markets. The results, in virtually every advanced capitalist country, were private housing booms, property speculation and office development of unheard-of magnitudes. The property boom was followed by an equally dramatic slump which at one stage threatened the whole of the world financial system. The perversity of the situation was noticed by many – substantial problems in their daily lives for most people amidst an orgy of property speculation – and many blamed the latter for the former. Landed property seemed to have come into stark contradiction with the rest of capitalist society, and land rent seemed to hold the key to understanding why.

The traditional political strategies of European social democratic and communist parties were, for different reasons, not equipped to meet the changing circumstances. More public expenditure and consensus proposals for reform of the housing and commercial property markets were insufficient, yet all that social democratic parties could offer, whilst the traditional production orientation of the communist parties offered no immediate strategy towards these issues at all. Left critiques of social democracy and attempts by intellectuals within communist parties, particularly in France and Italy, to change their parties' strategies in these fields were consequently of pressing political importance. One line of analysis was via the theory of rent. The idea behind such analysis was to unveil conflicts over land use and provide strategies that enabled those communist parties to gain political hegemony and thus strategic control over the diverse currents that made up urban protest movements.

Similar trends of radicalised social movements over ostensibly urban issues could be traced in the United States. The growth of a nonpolitically-aligned body of radicalised intellectuals concerned with urban issues coincided with, and was stimulated by, such optimistic political trends. A wide body of intellectual interest developed around urban issues, with high hopes for the simple application of Marxist analysis to urban problems, and heady expectations of their immediate political relevance and application. Enthusiasm for the rent question was substantial – it seemed the obvious way forward.

The period of renewed Marxist studies of land and rent in the early 1970s failed to develop effectively. From about 1975 there was a hiatus, a stagnation, which in our view has only started to be broken in the last few years – even though developments are still tentative and uneven between countries. This stagnant period can be described and to some extent explained.

The stagnation in rent studies reflects two kinds of forces: the narrow limitations of the analyses which had been made and the changing political context. Many of the writings and discussions of the early 1970s can be characterised, in terms of their method, as follows. First came the excitement of the rediscovery of Marx's treatment of rent. A great deal of exposition and paraphrasing of Marx's categories of agricultural rent took place. Differential rent, types I and II, monopoly rent and absolute rent became the key concepts structuring discourse. Some of the analysis was meticulous and accurate; some was essentially slapdash and eclectic. A common feature – and, we think, a major failing – was the attempt to base an urban analysis on the categories of agricultural rent, rather than on the methods Marx had used to generate them. This weakness often led to the simple equation of a category of rent with a particular kind of land-use conflict, even though the economic conditions that generated the usefulness of those categories of rent in agriculture could not be reproduced in the urban situation. Marx's discussion of absolute rent in agriculture, for example, was mechanically transposed into an urban absolute rent via the construction industry.

Since, by definition, absolute rents can arise only where there are variations in the organic composition of capital between branches of production, there was a tendency to assume that absolute rents could arise wherever there were highly labour-intensive industries. Since construction appeared to be such an industry, it became commonplace to see the urban development process, among other things, as the search for realisable absolute rents. We can now see that this constitutes a static and incomplete understanding of the idea of absolute rent.

Thus the attempts to use Marx's categories of rent in the early and mid-1970s had weaknesses from a theoretical point of view. But their central concern was to overthrow previously-accepted wisdoms and to formulate new political strategies. Direct attempts to relate the categories of rent to actual political and social conflicts led to some uncomfortable interpretations of empirical data, and disquiet over the validity of the whole approach. The rent analyses of the 1970s, nevertheless, fitted particularly well in their political contexts – fuelling the anti-monopoly thrust of French and Italian Communist thinking, sustaining the opposition of 'community' groups (real or contrived) to development capital on both sides of the Atlantic, and yielding critiques of the legislative efforts of the British Labour Party to nationalise (a very narrowly-defined category of) development profits.

In retrospect, this period of rent analysis did have some lasting benefits. The critique of previously held orthodoxies was a powerful one, even if the alternatives proposed were somewhat questionable. Moreover, the difficulties of a simple transposition of Marx's categories of agricultural rent to the urban situation are made most clear by attempts to try and do it. The political concern of most analyses undertaken also stopped the theory of rent from being restricted to narrow scholasticism. The political failing itself, furthermore, was one of trying to go from a limited segment of Marxist economic theory immediately into direct political strategy. The general impossibility of doing so has been learnt and well aired in later Marxist debates. This does not mean, however, that the analysis of rent has no place in the formulation of political strategies, just that it has to be seen as part of a complex set of determinations rather than as a means of sidestepping such necessary in-depth analysis.

The period of stagnation in the development of urban rent theory that emerged after the mid-1970s was not without theoretical advances. The housebuilding industries of Britain and France, for example, began to be subjected to more far-reaching study than they had received before, either from Marxists or from anyone else.

By the late 1970s, the political and economic context, to which interest in rent theory related, had changed dramatically. Left and social democratic parties were in retreat on many fronts. Economic crisis, deindustrialisation and the internationalisation of production were eroding the power basis of the old trade unions, and right-wing economic dogmas were gaining ideological and political ascendancy in many countries. This has not led to the demise of the urban problems mentioned earlier; if anything they have intensified. Instead, at present, political reaction to them is muted, or takes sudden, spontaneous forms such as the urban riots in Britain in 1981.

One major feature that stimulated earlier analyses of rent theory, property market booms and slumps, has continued to occur – for example, in the depths of recession in Britain in 1982, more new offices were built in central London than in the peak year of the early 1970s property boom. Substantial restructuring, however, has taken place in the property and housing sectors, so that firms are less vulnerable to, though not immune from, spectacular bankruptcy. The economic mechanisms of these sectors consequently have less of a high media profile than previously. This has contributed to the period of relative political quiescence.

But, as the 'urban' problems of earlier years have intensified rather than disappeared, it is to be expected that they will return to high prominence on the political agenda at some time in the future. The depth of the changes that are taking place in capitalist societies as a whole, and in their urban development processes in particular, moreover, make it increasingly unlikely that the old political slogans can be dusted off and used again. So, even though there might be temporary political quiescence, a space has been opened for the development of radically different new political ideas on urban problems. It is possible that such potential opportunities for proposing new political strategies has not existed for thirty years or more.

The political retrenchment of the late 1970s appears to have coincided with some much more effective and thorough intellectual work, building on the lessons and advances of the past. In particular, it has been recognised by many that political advance on the left can only be achieved by building broad alliances between social movements operating across a series of diverse issues. Attempts to contain them all under the control of one political party are recognised to be doomed to failure. Moreover, the lesson of not being able to read off political actions directly from economic circumstances has been painfully learnt and relearnt. Social movements are becoming more clearly understood as not being static entities which can easily be drawn into and subordinated to the principal contradictions between capital and labour. The analysis of landed property consequently has taken on new dimensions.

The growth of radical ideas in economics, geography and urban studies in general has also been very important: generating work on many aspects of the process of capital accumulation; on international capital movements and the international division of labour; on the dynamics of accumulation in specific industries; on the relationship between the drive for accumulation and the technology and labour process in construction, and on the spatial dimensions of the restructuring of capital. Thus we are again in a period where productive and fruitful work is being done in landed property, rent and urban development.

Current Themes of the Analysis of Landed Property and Urban Rent

The papers in this book reflect the range of the current debate going on over rent. Not all of them agree with each other, whilst each deals with a separate empirical or theoretical topic. Yet we feel there is a common theme running through the papers which reflects broad agreement on the basis on which the analysis of rent should proceed, and on its relation to problems as diverse as minerals extraction, office development, housing provision and land-use planning. In this section we summarise what we feel are the key theoretical elements in contemporary Marxist analysis of rent.

The central argument of the book is that struggles over land rent and land prices are important components of the social processes of land development. As such they form part of the struggle surrounding the process of capital accumulation at large. One cannot first analyse the accumulation process in a spatial vacuum and then make a separate analysis of location, land and so forth. The exploitation of rental opportunities, of differences between places, between labour forces and between jurisdictions is an integral part of accumulation.

The urban development process is an exceedingly complex one. Part of the complexity lies in the diversity of forms of landownership and of landed assets; part lies in the heterogeneity of the construction and development industries with their associated labour processes and technologies. Thus struggles over the built environment cannot be analysed in isolation from the contexts in which they take place. In other words, the influence of rent (on such things as the kinds of housing provided, the organisation and production methods of the building industry, the interventions of the state through planning) is very specific – historically and nationally. The form and impact of rent depends on the social relations and the institutional framework in which land development takes place. The balance of social forces will determine the nature and form of development. Thus consideration of the pressures flowing from each of the social agents in the process is essential to an understanding of the actual built environment which results – e.g. the office development cycle, the relative growth of housing tenures, the impact of land-use planning. The attempts by some agents to appropriate land rents, and the attempts of others to avoid paying them, are crucial features of the development process. Thus land rents cannot be understood except in the specific context where they arise, and conversely we cannot expect to understand the urbanisation process as a whole without looking at the part played in it by rent.

A central theme running through the papers is that the analysis of rent relations helps to bring out the historical specificity of the accumulation and development processes in which landed property intervenes. While certain general tendencies in urban development can be seen in most advanced capitalist countries, such as the contemporary growth of owner occupation, more detailed investigation reveals wide divergences as well. A key component of these divergences is in the means whereby landed property in each country tries to appropriate rent – and its success in doing so. The comparison of recent trends in British and French owner-occupation brings this point out well.

The processes involving the realisation of rents and development surpluses are thus seen as complex and geographically specific. Equally important they have to be seen as dynamic, since one of the central features of the private ownership of land is the potential barrier it poses to flows of capital – between locations, between branches of production and so on. In a fast-changing world the housing speculator and the investment institution know that their job is to maximise the growth of their share of surplus value (or, at times of crisis, to slow its rate of decline): the application of theory must at least match the dynamism of the world to which it is applied.

The meaning of the word 'land' is itself complex. It refers to a concept relating to the earth, to locations, to jurisdictions and, above all, to ownership. We have to acknowledge that, as concepts of ownership – and legal definitions of ownership rights – have changed, so the meaning of the word 'land' has changed. It embodies a great deal of history and thus its use in discussion can easily import an unconscious conservatism into the debate. A society in which land assets were not traded, for instance, would soon develop new meaning for the word 'land'.

The historically contingent meaning of the concept 'land' arises because of the historically specific nature of class societies. No universal theory of rent exists; rent is not a payment to an inanimate thing called land but to a category of social agents defined by their property rights in land, commonly known in the literature as 'landed property'. Rent is the economic payment to landed property resulting from its monopoly ownership of land and/or the buildings standing on it. Strictly, there is no such thing as a 'theory of rent', instead there are theories of the economic place of landed property within a given type of class society (capitalist, etc). Rent is only one element of the economic role of landed property, and sole emphasis on it can give a distorted impression of the problems and contradictions generated by the existence of landed property.

However, the idea that theories of rent are equivalent to, and sufficient for an understanding of the economic role of landed property is so ingrained within popular and academic discourse that reference to the study of land rent carries with it an ambiguity. As the following chapters are concerned to make their arguments as jargon-free and accessible as possible, this ambiguity over 'rent theory' is reproduced here. But, the context makes it clear that it is landed property that is being discussed, not some socially neutral thing called 'rent'.

Finally, it is important to remember that the class situation of landed property cannot be deduced in isolation from a concrete understanding of the class structure of the society in question. The following chapters are almost entirely concerned with capitalist societies and, hence, the situation of landed property within them. Even so the class location of landed property can be highly varied. Traditional notions of an independent landed class, in particular, do not accurately reflect the complexities of landownership in many advanced capitalist societies. One noticeable trend of the past 25 years, as a number of the following chapters point out, is for landed property to become just another sphere for the investment of capital. In such cases, landed property becomes indistinguishable from capital itself. This does not remove the potential contradictions of private landownership for the accumulation of capital, instead it transforms the position of landed property within wider political and social struggles. This theme is taken up in more detail in the last section of the book.

Organisation of the Book

The book is arranged as a series of interconnected readings in four sections, each dealing with specific topics (of which this introduction is the first). The second section deals with the practical effects of land rent on development processes. The third presents a range of viewpoints on the Marxist theory of ground rent itself. The concluding section raises the political issues associated with land and urban development in Europe and considers ways forward for research and for political debate. The book ends with a selective bibliography of the key texts which readers may find helpful in penetrating further into the subject.

Synopsis of the Following Chapters

The remainder of this Introduction consists of a summary of the subsequent chapters. This is designed to help those new to the field select the most appropriate point of entry to the debates. In the process it offers an overview of the book which others may find useful as well.

Part Two

Chapters 2 to 6 comprise Part Two of the book, presenting a number of empirical perspectives on the social relations of land development in European countries.

Chapter 2 Prices, Profits and Rents in Residential Development: France 1960-80 by Christian Topalov. The chapter presents a first view of the social relations of land development. It reports an analysis of the changes in costs and profits of new housing construction in Paris and the provinces, concentrating on the struggles over the division of surplus profits between developers, builders, financial interests, materials suppliers and landowners. This struggle is viewed both in terms of the division of the total surplus and the rates of profit on capital advanced by those concerned. The data are examined in relation to changes in the French economy and the results are important both methodologically and in substance.

The analysis suggests a periodisation of residential development in the past 25 years as follows. Between 1960 and 1965 developers, concentrating on luxury sub-markets at a time of stable building costs, were able to realise growing developmental profits while selling prices rose. Speculative land-trading forced up the costs of raw land creating a rapid inflationary spiral. Inflation finally overstrained the ability of the financial system to fund the purchases. The resultant crisis of 1965 introduced a period in which development profits fell back to lower levels, while debt-charges on outstanding loans and on work-in-progress came to take larger parts of the total sale price – and house prices themselves fell in real terms. An increasing dominance by the banks was partially hidden by relatively low interest rates. In the boom of 1968–74 the system newly dominated by monopoly capital – especially by banks – grew. The 1974 crisis again saw a rapid fall in output, but the selling prices of dwellings continued to rise – albeit more slowly. Falling sales pushed financing costs further upwards and saw the virtual demise of the developers, who were replaced by financial institutions on the one hand and by petit bourgeois forms of individualised house building on the other. The bank-dominated system is itself in crisis now but this time it is harder for prices to fall: land and development profits are not there to be squeezed.

In terms of method the paper proposes and uses an empirical analysis of recent development history, informed by – but not pre-structured around – a theoretical understanding of the struggle for development surpluses.

Chapter 3 Rent and Housing Development Processes in Europe: Some Hypotheses from a Comparative Analysis by Marino Folin. Since the end of the Second World War more dwellings have been built in Europe than ever before. Throughout Europe housing problems have been considered primarily in terms of physical shortages and solutions have been found in the construction of new dwellings. Behind this political decision there was a conviction that the production of housing could have positive effects on the accumulation process because of the proposed nature of 'investment' in it. During the 1970s new housing output began to decline. The official explanations of this phenomenon have been couched in terms either of a natural decline due to disappearing shortages and/or of macroeconomic problems, town planning constraints or other features external to the nature of the market itself. No analyses are made either of the historical development of housing needs or of the peculiar nature of investment in housing. (Housing cannot be treated like a machine tool, a vehicle or any other means of production.) It is argued that trends in housing production can only be understood if we look at the diverse forms of housing provision – at the variations defined in terms of the distinct social agents involved in their production and exchange. Housing development processes are not simply the 'carrying out of building . . . in, on, under or over . . . land' as suggested in the British town planning acts from 1947 onwards, but a valorisation process which can be understood only through the complexity of social production.

The author examines changes in housing production in a number of European countries since the War (France, Great Britain, the Netherlands, West Germany and Sweden). He considers them with reference to the different kinds of development agencies involved in housing provision – private as well as public. Each form of development process defines specific and distinctive relationships between the actors involved in it and so has distinct effects on the creation and appropriation of value in housing production. The paper describes and analyses the nature of each type of developer; finally it suggests some hypotheses to explain the decline in housing output which has been common to all the countries considered.

Chapter 4 Land Rent and the Construction Industry by Michael Ball. In this chapter Ball first explains in detail why the categories of rent Marx derived from the analysis of nineteenth-century agriculture cannot simply be transposed to the contemporary construction industry. Marx himself argued that the price of land may be determined by many 'fortuitous combinations' of forces and his own examination of agriculture was essentially an elucidation of the forces obtaining there. Two central features of that account were (i) because farming produces a more-or-less homogeneous product at diverse locations the prevailing market prices were determined by the prices of production on the least favoured farms; and (ii) that it was demonstrable that non-zero rents were charged even for marginal land – so that the formulation of the idea of absolute rent could be based on a tangible barrier to capital inflow, not just on a circular argument about low levels of organic composition of agricultural capital. These two specific features of agriculture are not paralleled in modern urban development and Ball summarises the kind of analysis which must be made if contemporary construction is to be understood in its relation to land.

In contract building the role of rent can only be indirect – via the forces acting on the builder's client – since the contractor is insulated from struggles over rent. By contrast speculative builders are deeply engaged in the competition for land development profits, though in ways which are locally and historically specific. In early nineteenth-century British house building land rent and the prevailing forms of land ownership had positive effects on the flow of capital into construction. In the years before the First World War, however, rent relations appear to have had a damaging effect on speculative building for rent and to have helped precipitate the crisis of that form of provision. The analysis shows how, more recently, rent relations have had powerful effects upon the scale and character of speculative house building and upon the technologies and working conditions which have prevailed.

Chapter 5 The Law of the Land: Property Rights and Town Planning in Modern Britain by Michael McMahon. The author argues that the history of town planning is frequently written from a wholly erroneous point of view: first a free market in land is assumed to exist; increasingly severe contradictions then develop (congestion, pollution, lack of open space, ribbon development and so on); state intervention is thus needed and justified to resolve them. The ability of the modified market system to deliver the 'optimal' pattern of urban development then becomes the issue. The present paper shows that in fact no such 'free' market in land rights existed in Britain and that the central role of the state in the early history of planning was to oversee and orchestrate the emergence of a market with at least some elements of 'freedom'. The study traces the mounting pressures on the state during the nineteenth century to resolve the growing contradictions between the entailed ownerships of the landed class and the needs of capital for free access to land for new urbanisation and for railway and infrastructure development. In the British context at least, the key historical question is not how a developed market generates a need for planning, but what were the social forces underlying the shift from the rigid monopoly of former landed property to an age in which property rights were both publicly administered through planning and the objects of speculative investment.

Chapter 6 Land, Capital and the British Coal Industry Prior to World War II by Ben Fine. Drawing on a larger study of the industry, Fine shows how the forms of ownership of mineral and land rights first facilitated the development of the industry (in contrast to conditions elsewhere in Europe) but later increasingly became a barrier to investment and a source of mounting contradiction – only resolved by the nationalisation of mineral rights. The large landed estates which (with exceptions) prevailed in Britain, provided ideal conditions for the rapid growth of mining in the early nineteenth century – both for the flow of capital into underground works on an efficient scale and for surface developments for transporting the coal and reproducing the work-force. Increasingly, however, the boundaries of coal ownerships (which corresponded to surface estates) obstructed the growth of mines, and especially their rationalisation. A state-supported cartel was required to maintain an adequate rate of profit in these conditions, each mine being allocated a quota. Despite various official inquiries, the problems of landownership remained unresolved, with the royalty-owners and mining capitalists divided in the face of an increasingly organised labour force.

The study shows how rents of various kinds were appropriated. In particular it is found that the royalty system depressed and distorted investment in large-scale mechanisation and that the realisation of the extra rents associated with such investment (what Marx termed differential rent II) was intricately tied up with the vagaries of ownership. As the national pressures for re-structuring mounted later in the century (and were largely frustrated) conditions existed for the realisation of absolute rents – and the study shows how they were actually appropriated. This analysis is able to indicate, as it goes along, how forms of land ownership and rent influenced the labour process in the mines themselves – the level of mechanisation, the forms of supervision and the working and safety conditions.

Part Three

Chapters 7, 8 and 9 confront Marx's theoretical work on rent directly.

Chapter 7 A Marxist Approach to Urban Ground Rent: The Case of France by Alain Lipietz. This chapter presents a distillation of the author's *Le Tribut Foncier Urbain*, of 1974, never published in English, and draws also on some more recent work. A brief epistemological critique of the neo-classical orthodoxy in Anglo-Saxon and French literature flows into an argument that the 'social division of space' should be regarded as a product of history analogous to (and reflecting) the social division of labour. Within this framework the bulk of the paper develops an analysis of the sources and meaning of the payments which developers must make for the land they develop – payments which Lipietz designates as 'tribute'. The analysis concentrates on the source of surplus profits in housing development, contrasting the approaches dubbed 'rent *à la* Marx' and 'rent *à la* Engels': the former sees the prevailing average levels of rent (and tribute) as a redistribution of surpluses generated within the land-using industry (e.g. construction); the latter sees it as part of the general redistribution of the social surplus. The political and analytical implications of the two views are discussed. An important feature of the analysis is the contrast between the appropriation of what are classified as 'exogenous' rent opportunities – where a developer essentially pockets the potential surplus profit which a site offers by virtue of its position in the social division of space – and 'endogenous' rent which is created where the developer (usually with state support) transforms the social division of space to create a new opportunity.

Chapter 8 Capitalist Urban Rent by Ambroise Gravejet. Gravejat sets out to demonstrate the historical specificity of capitalist urban rent, and to simplify our use of the concept in contemporary urban analysis, through a review of certain preceding rent forms. The general conditions for the realisation of ground rents are outlined, particularly those associated with increases in investment. While a general concept of rent may be valid for any social situation, only concrete examination of specific societies can lead to an understanding of the level and composition of rents and of the forms through which they are appropriated.

In ancient Rome, after the end of the Republic, and well into the Empire, we know that some 80 per cent of the population lived in very bad housing conditions, in overcrowded rented dwellings. This housing appears to have been built of poor materials and with slave labour; its construction costs probably accounted for less than half its rental value; thus ground rents were effectively being realised on a grand scale. Tenants were paying for the political benefits of citizenship, for infrastructure and cultural benefits, and were making their payments out of the surpluses from slave-cultivation of their rural lands. These rents were therefore transfers from one branch of production to another and were based on the particular form of Roman imperialism. An analysis of rents in mediaeval French towns shows how high rent levels for housing effectively formed a levy upon the merchant activities associated with the towns' rights to hold markets and fairs: here again we find agricultural surplus labour as the source of urban rents. In nineteenth-century France housing for the bourgeoisie was built to yield long-term profits while working-class housing developers worked on very short repayment periods. Capital was flowing between branches of production in a competitive pursuit of profit and the high levels of rent in working-class housing had their origins primarily in the surplus value generated within the construction process. In conclusion the paper calls for more attention to close historical study of the development process.

Chapter 9 Marxian Categories and the Determination of Land Prices by Agostino Nardocci. The core of the paper is a detailed examination of the three distinct formulations of rent to be found in Marx's work. Two presentations are in *Theories of Surplus Value* and a third in *Capital, III*. The interrelationship of these three versions is explored and their distinctive analytical and political implications are exposed. The author suggests that confusion about what Marx wrote may have contributed to the recent problems of applying Marx's rent categories to modern urban development and housing problems.

Part Four

The two final chapters of the book deal with the links between the theorisation of rent and the political strategies adopted by the left in Europe.

Chapter 10 Theory of Urban Rent and the Working-class Movement: The Case of Italy by Vincenzo Bentivegna. This chapter examines the relationship between theory (of rent and landownership) and the actual praxis of the working-class movement in Italy. The post-war strategy of the Italian left presupposed an essentially distributional role for rent. Landed property was seen as an anachronistic and inessential feature of modern capitalism, responsible for certain urban problems and influences on construction but mainly important in redistributing the mass of surplus value – whose magnitude was quite independently determined. That view underlay the consensus in which the Italian left settled for certain limited gains (provision of state housing, rational planning of urban development and social infrastructure, state involvement in the redistribution of wealth and income). But this consensus did not include any challenge to the private ownership of the means of production nor to the private control of building production. Where rent is seen as merely redistributional, rent struggles appear quite detached from struggles over wages and production; it seems possible for an alliance of workers and capitalists to isolate and subordinate the landed interest as a common enemy. We can now see that such a strategy (despite its successes) was short-sighted and based on poor theory.

By contrast we are now coming to see rent not just as distributional but as part of the whole process of production and realisation of surplus. The appropriation of rents (increasingly by financial capital, not old-style landowners) can play an influential role in determining what gets produced, where the production takes place and in forming the actual conditions of construction work. The whole momentum and direction of the accumulation process is affected by, and in turn affects, the opportunities for the realisation of rents. Private land ownership and urban rents are real barriers to the development of the productive forces, but increasingly this is a contradiction within capital, not between capital and a landed class. In this analysis there is no space for a redistributional alliance between capital and labour against a landed interest: the thrust of the struggle on urban questions comes to be part of the larger and indivisible struggle against the capitalist mode of production as a whole. The character and limitations of accumulation in the construction sector, and the transformation and use of scarce land resources become integrated in the critique of capitalism as a whole.

Chapter 11 Planning and the Land Market: Problems, Prospects and Strategy by Michael Edwards. This final chapter reviews the contemporary planning and land-development problems confronting the labour movement, concentrating on the British experience but with parallels drawn elsewhere. Particular importance is attached to the changing structure of financial capital and its behaviour in the face of profitability crises in the UK. The growth of housing owner-occupation and of funded pension schemes gives large parts of the working class a new and direct interest in the continued valorisation of landed assets and is associated with an increased blurring of the relations between state agencies and private development capital. The limited freedom from financial pressures which land-use-planning systems had sometimes secured has come under great pressure and the actions of central and local government bodies in Britain are increasingly geared to the dictates of private appropriation of the social surplus. The interaction with the methods and the products of the building industry are indicated and it is suggested that the existence of absolute rents and related barriers to investment obscures high rates of exploitation in a number of industries where technical change is frustrated. The paper concludes by urging that land-nationalisation campaigns in European countries should never again be fought in isolation from attempts to transform the systems of financial capital and of production itself.

Note

1. This historical survey, for clarity and brevity, avoids reference to particular articles and books published over the past fifteen years. A bibliography of the key texts is given at the end of the book.

PART TWO:
THE SOCIAL RELATIONS OF LAND DEVELOPMENT: A EUROPEAN PERSPECTIVE

2 PRICES, PROFITS AND RENTS IN RESIDENTIAL DEVELOPMENT: FRANCE 1960–80

Christian Topalov

Introduction

My starting point is one of the classic ideas in Marx's epistemology, that price is not an intrinsic property of an object or of merchandise but rather an expression of social relations in the sphere of the market and of money. I do not intend to discuss this proposition but shall concentrate on implementing it, on bringing it into play empirically in the area which concerns us here.

Before explaining the empirical methodology, however, I shall make a brief summary of the theoretical context and framework in which it is set.[1]

The Formation and Division of Development Profits: a Theoretical Framework

The competition between capitals in their search for maximum profit brings about flows of value and of capital among the various branches of production and tends to create an average rate of profit. The capitals engaged in the various parts of the property sector cannot avoid this general tendency.

However, the characteristics of the relations of production and circulation in the sector are so particular that some very specific flows of value develop between it and the rest of the economy and upset the general tendency to harmonisation of rates of profit. In fact the excess profits which are shielded from this harmonisation process end up in the property sector mainly through the fact that, in housing, the price is above the price of production. On the one hand a proportion of the surplus value created in building projects is retained in the building sector (over and above the normal rate of profit); on the other hand the property sector draws heavily on the surplus value created in other sectors. This double flow has a depressing effect on the generally-prevailing rate of profit elsewhere.

The retention of this surplus value leads to the raising of the generally-ruling prices of new dwellings above their average prices of production. The retention flows from two distinct processes – which lead to the formation of the only two components determining housing prices. Absolute rent from buildings is a result of the effective resistance of the owners of developable sites to changes of use. This resistance has the effect, on the one hand, of transferring to landed property the excess profits created in construction by 'uncoupling' the price of developable sites (even the least attractive ones) from the price of farm land. On the other hand variations between locations in the economics of development lead to the appropriation of surplus profits in the form of differential rents. These rents together prevent the prevailing prices of dwellings from establishing themselves at the average social cost of production and tend to sustain them at the level characteristic of the least favourable production conditions.

Added to these obstacles to the harmonisation of profits is the drain imposed by the development sector on the surplus value produced in the rest of the economy. Some of the current production of housing is sold at prices not related to prevailing prices (even the inflated prices described above) but to demand conditions alone. These monopoly prices result from the lack of substitutes for the most prestigious locations and from the strong purchasing power of the higher strata of society. They engender surplus profits which can be transformed into rents, rents due to social segregation. In contrast to differential and absolute rents, these monopoly rents do not influence the price of housing: they are determined by it. Essentially the drain which results can be seen as a tax on the higher social strata, constituting a kind of consumption of the surplus value already distributed among these rich households. It thus has no effect on the general rate of profit.

The picture changes, however, if we look at the whole housing market, not just at the market for new dwellings. One part of the stock is the subject of exchange processes – for resale or rental. Prices in this standing stock are at a level broadly determined by the prevailing prices for new dwellings. The exceptions are the sub-markets in very special and unusual houses where the short supply generates a monopoly price. The difference between the prices for the standing stock and the prices which would have resulted from the repayment of the original investments (with interest) is the essence of housing rent. The payment of these rents (or purchase prices) by a substantial mass of households represents another drain to the property sector from the rest of the economy. These rents have a direct effect on the purchasing power of the employed classes and in some circumstances can influence the levels of wages and of profits. To a significant extent the chains of buying and selling in the housing market ultimately feed the demand for new housing, especially that segment of demand met by the developers and which sustains monopoly rents associated with segregation.

The property sector thus operates both by retaining surplus profits created internally and by diverting them from elsewhere – and both operations occur through the prices of houses. These surplus profits, their incorporation in prices and their transformation into rents are all the result of social relations. To analyse them is thus to study the division of surplus profits among the competing agents who make up the property sector.

The respective levels of house prices and of the elements necessary to production – labour, components and materials and fixed capital – determine the total amount of profit available for division, in varying proportions through time, among the agents. We can approximate these profits via the measurable components of the prices of new houses.

Price and Cost for New Housing: Problems of Method

Let us be precise about the aims and the limitations of the analysis which follows. The aim is an examination of the formation and the distribution of profits at the heart of the system of private housing development and the long-term changes in this distribution. The data are, on the one hand, for the prices of dwellings sold by the developers; on the other hand for the structure of the developers' costs. These data only became available during the 1960s and thus set the time-span of the study.

I am going to argue that the quantitative evidence on prices and costs can be taken as indicators of the structural changes at the heart of the system of agents being examined. Two questions are posed. What changes have there been in the total mass of surplus value appropriated by the property sector? And how does the distribution take place between developers, landowners, building firms and financial institutions? This formulation raises some delicate problems.

The first is linked to the fact that the data naturally measure only the price form of the fluctuations in value. In view of our aims this problem does not trouble me much. There is no need to examine the role of the development sector within the totality of social production, the determination of the prevailing rate of profit, the accumulation process and its crisis. Others have done that work, which does entail working in value terms to establish where value is created and how it circulates (Secchi 1972; Harvey 1982). My aim is more limited because I'm looking at a single sector and a particular set of agents. It seems to me acceptable in this context to work in terms of costs and prices rather than in terms of value and surplus value. It is more serious that we cannot identify the price of production – only the selling prices. Thus we cannot distinguish, within that part of the final price which goes to each of the agents, how much is due to capital consumption, how much to normal profit and how much to surplus profit in its diverse forms. It is in intrinsic problem of the price form that real transfers of surplus value take place between agents but this always appears as an intrinsic part of the cost of the commodities being traded. Thus real social relations remain obscured – even from the participants themselves. The interpretation of the available data requires the formulation of hypotheses, some of which are rather hazardous. It is in fact the examination of the relative changes over time in the make-up of prices which gives the hypotheses a certain plausibility.

A further difficulty is in setting boundaries to the study. My focus is the central agent in the process of transformation of the built environment – the developer – and I study his relationships with other participants. I am not trying to make any comprehensive study of the determination of profits in the construction industry or its tributary industries, nor in the banking and financial sector. Any other choice would require a study of inter-industry relations and of the financial system in their entirety. Clearly the development industry is enmeshed in these global relationships but their analysis is a separate research project.

Even so, one of the two questions concerns changes in the global mass of surplus value appropriated by the private development system (that is, the mass which stands to be divided among the participants). The problem would have scarcely any meaning if the price of a house were determined simply by its price of production. This is not the case, for two reasons. First because of the phenomenon of rent: localised surplus profits, whether they determine housing prices or are determined by them, are formed independently of prices of production and express themselves as rents. Secondly because the financing of development involves two kinds of capital: capital valued in relation to an average (or higher) rate of profit and capital which has been devalorised – in France, essentially, by state intervention. This combination is reflected in the prevailing prices of houses and allows for the incorporation of some surplus profit in those prices. Although less stable than those which stem from rents, these surplus profits form a part of the global mass for distribution.

These two characteristics of the residential development system – rents and the combination of capitals – are fundamentally social relationships and are reflected in the price levels for new houses. Granted, these prices contain some elements which should be analysed not in terms of the division of surplus profits but in terms of the production of value. It is, however, generally agreed that over the last 20 years, and especially in the private housing developers' sub-market, there have been insufficient technical changes to produce any major growth of productivity, any 'value revolution' (*révolution de la valeur*). If this is so, it is legitimate to consider the changes in selling prices of dwellings (expressed in constant-price terms) as a convenient dynamic indicator of the surplus value retained within or transferred to the development sector.

The second question addressed to the data is about the division among the agents of this surplus value which constitutes surplus profit. The structure of sale prices of new houses changes over time (see Table 2.1). The relative movements of the components of this final price can be considered as indicators of the division of surplus profit among the agents. This proposition is now justified briefly.

The term 'construction cost' refers to the market price of the building work. In as much as the standard of new houses sold has changed little over the period, this cost (per square metre of usable space) reflects changes in prices, not changes in use values. We shall treat it here as broadly regulated by the price of production of buildings. As we shall see, this is a simplification because the average price of construction work can incorporate elements which end up as the surplus profits of building firms, building materials producers and banks. But it is a legitimate simplification because this part of the construction sector is both capitalist and competitive. Private residential developers tend to use neither the very biggest construction firms nor the very smallest, let alone the artisans (Topalov 1974; Pottier and Rogalski 1977; Combes and Imbert 1978). The remaining empirical problem is that the available data on building prices tell us nothing about the profitability of building firms. The study of the share-out of surplus profit is thus essentially limited to the forms it takes in the non-productive part of the development sector.

The other forms of surplus profit are in direct opposition because the agents concerned are competing with each other for their shares.

When the developer sells his finished building he initially appropriates the whole of the surplus profit; he then distributes all or part of it to the landowner from whom he bought the site and to the bank which lent the working capital. This carve-up will operate in variable proportions, dependent on the general conditions which prevail and on the specific relations between the agents. How can we account for this in quantified terms?

The 'development margin' reflects essentially the profit on developers' capital. It is no doubt necessary to distinguish between the earnings of the developers themselves – in respect of their work done – and the genuine development profit which is a return to the capital involved in financing the job. In fact the development margin also finances the expenses of development – management and marketing – which are neither negligible nor constant through time. One can, however, assume that the net profit on development capital moves in line with the movement of the whole – which comprises this net margin and the management and marketing expenses. Strictly speaking this net profit from development consists of two parts (utterly inseparable in practice): the profit at the prevailing average level and a part of the locally-obtainable surplus profit. This distinction is not just an abstraction: if the scale of profitability falls so much that average profits are eaten into, capital is liable to leave the sector altogether. If, in contrast, profitability rises and can be retained by the developers there will be a flow of capital into the sector (and competition from the other agents for a share of it).

The development margin is in fact in direct competition with two other elements of 'cost'. The 'financing cost' represents the pre-emption by loan capital of some of the surplus. While this capital is provided by banks, mainly as short-term credit, the arrangement can have a real impact on the profits of the project and can augment the profits (or the surplus profits) of the banks.

The 'site price' or site cost is made up largely of the transfer to the landowner of all or part of the localised surplus profit created by the development. It is thus in direct tension and competition with the two other elements of surplus profit just described.

If this is so, we can see that the accountancy of the operation seems to show the exact reverse of the real social process. Things come to seem like costs of components of the final product which are objectively forms of subdivision of the whole surplus value appropriated by the system and shared out among the participants: development capital, banking capital and landed property. The empirical study shows just

Table 2.1: Nomenclature and Definitions of Categories for the Analysis of the Composition of New House Prices

Individual categories	Aggregated categories					
1. Site price (paid to seller) [*prix du terrain - prix net payé au vendeur*]	13. Site acquisition cost [*prix de revient du terrain*]					
2. Site transaction costs (survey, tax, legal; disturbance payments, demolition and eviction costs) [*Frais d'aquisition . . .*]		15. Cost of developed sites [*charge foncière*]				
3. Spending on network services [*Dépenses de voirie et réseaux divers*]	14. Site development expenditure [*Défenses annexes au terrain*]					
4. Contributions to public services (local charges for infrastructure, cost of 'planning gain', free land transfers etc) [*taxes locale d'équipement ou participation, taxe dépassement do COS, cessions gratuites de terrains ou travaux gratuites, etc.*]			17. Technical development expenditure [*Prix de revient 'technique'*]			
5. Site preparation costs (levelling, landscaping), garages and parking spaces [*Dépenses d'aménagement du terrain — terrassements, clôtures, espaces verts — de garages, parkings*]				19. Total development expenditure [*Prix de revient*]	20. Selling price [*Prix de vente*]	
6. Building expenditure (incl. foundations) [*Dépenses de bâtiment (y compris fondations spéciales)*]		16. Construction cost [*Coût de construction*]				
7. Fees related to building (architect, QS, planning studies etc) [*architecte, métreur, bureau d'études*]						
8. Financing costs [*frais financiers*]			18. Overhead expenses [*frais annexes*]			21. Gross development surplus [*marge réelle de promotion'*]
9. Any residual VAT [*TVA résiduelle*]						
10. Developer's management [*frais de gestion*]						
11. Marketing [*frais de commercialisation*]						
12. Net development surplus (or apparent development surplus: return on development capital, pre-tax) [*Marge net de promotion (rémuneration du capital de promotion, brute d'impôt) ou 'marge apparente de promotion'*]						

Note: The terms 'cost', 'price', 'expenditure' etc. are used more or less interchangeably here (as in the French). The conceptual problems of the terminology are discussed at length in the text. (Eds.)

how consistent this theoretically-derived perspective is with the actual data.

One final comment on the technical problems of the data.[2] They are often discontinuous in time and hard to assemble as long, consistent series – especially for some key variables: development capital, banking capital and landed property. It is often necessary to rely on disparate sources, parallels from other fields or on data whose range is limited. Finally, in studies of house prices not only does nomenclature vary but even the definition of the observed price can change: sometimes the 'sale price' excludes the overhead expenses and sometimes it is a genuine sale price in the terms defined in Table 2.1.[3] These heterogeneities are a problem for rigorous analysis, but they are not just the results of bureaucratic irrationalities and inconsistency. They show once more that statistical 'data' are a social product. The periodic shifts in perspective concerning house prices among those concerned – and especially among public officials – tend to lead to changes in the statistical arrangements. These changes are themselves an index of new social realities. Although an active policy of public intervention in financing and construction dates back to the mid-1950s in France, it was only in the mid-1960s that an appropriate statistical apparatus was created. This was the time when property development experienced a real boom, when it was released from the price controls associated with state-subsidised loans and produced a strong impetus to inflation. The practical problems the state experienced in that period prompted a mass of research on prices and costs. But the climate changed again: it is not by chance that the excitement of land-market research in the mid-1960s has given way to a new focus on 'overhead expenses' since 1975.[4]

The Price of Dwellings and the Price of Building

A dwelling is not like any other kind of merchandise. It is a building, the product of work done in the construction industry. It is also a property, a building on a site, with a selling price which includes surplus profits and rents. We shall show that the trends in selling prices and in building costs are autonomous: autonomous but not fully independent since the law of value is at work in this industry. It is mainly because of the familiar phenomenon of slow productivity growth in construction that the real price of new dwellings has such a sustained upward trend, pulling the rest of the market up with it. But price movements reflect

Figure 2.1: Selling Prices of New Dwellings, 1960-80

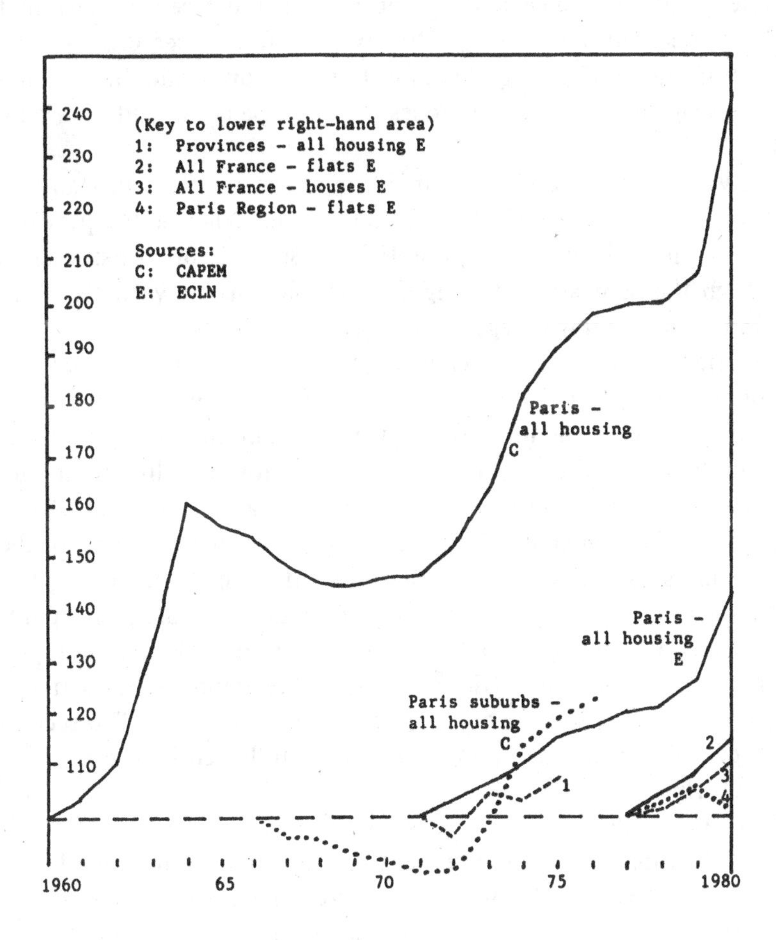

Sources: CAPEM and ECLN.

other factors too, which change with the balance between the agents concerned in development. This is the space – between costs and prices – where banking and development profits (and surplus profits) arise.

The Sale Price of New Dwellings

The available evidence is in broad agreement on the price movements for housing sold by developers since the early 1960s (see Figure 2.1). Three fairly distinct periods are apparent.

Between 1961 and 1964 prices rose at a speed which was probably unprecedented since the War. Between 1965 and 1971, however, real prices fell and then stagnated – contrary to the received view that property prices always rise. In Paris even money prices dropped between 1965 and 1967, rising thereafter but at a slower rate than general inflation. Only in 1970 did real prices begin to rise again, and then only slowly.[5]

A new period of real growth in prices began in 1973, though at a less dramatic speed than in the previous boom. Whereas the previous boom was violently halted by the 1965 crisis, this price rise carried on through the depression starting in 1975, slowing between 1977 and 1978 and then speeding up again, especially in Paris.

The price of housing thus changes in response to the conjunctural conditions, though in ways which vary from one cycle to the next. In the 1961-8 cycle, the increase was very rapid in the boom and prices fell in the crisis. In the next crisis prices rose slowly in the boom and carried on rising in the crisis. We shall look for an explanation of this difference in the changing relations among the agents. The core of the explanation is as follows. The first rapid inflation reflected the fast growth of property development in a form dominated by the petit-bourgeoisie. The subsequent fall in the relative price of housing represented the crisis for that particular form of development and a transition to forms of development dominated by monopolies. The second growth period reflects the rapid expansion – followed by a crisis – of this new form.

If there is really this link between the price movements and the structural features of the development system then the periods should show up quite distinctly in the components which go to make up the prices. Thus we can form some hypotheses to address to the data in order to verify this interpretation. (See Topalov (1974) for a fuller treatment.)

The price rise of the 1961-8 boom is mainly a product of an imbalance between the overall supply of housing and effective demand. The very poor response of supply is doubtless the lasting legacy of the technical characteristics of the building industry, especially in the production of blocks of flats. This rigidity was specially marked in the early 1960s. Building firms were numerous and small with very low rates of accumulation. The development business, likewise highly atomised, had no large capital funds of its own, nor access to banking capital and was very much dependent on the inherited wealth of private entrepreneurs. Demand, on the other hand, burgeoned. The factor which tipped the market into disequilibrium was undoubtedly the return to France of the expatriates from Algeria.[6]

These households increased the demand for owner-occupation while the capitals fleeing the erstwhile colony added to the flow of investment. But other factors were at the root of this sustained demand. It was a period when households, even in quite low social strata, were sinking their inheritances in owner-occupation. Housing loans were determined much more by people's pre-existing savings than by their actual or expected incomes. Initial deposits were very important and thus the limit placed by income on borrowing capacity was not very strict – the opposite of what we shall find in the 1970s. Clearly there were inheritances available to bring about this change, and they were to feed the boom throughout its duration. Once their effects (on increased production and on increased prices) had been spent the market came to depend much more on the availability of credit and on its cost. Before and after the reform of development credit in 1963 developers had been deserting the sectors aided by state finance and had rediscovered the freedom to raise prices. In that period the banks were unable to grant full long-term loans to purchasers and insisted on large deposits. The growth of demand in these conditions was only possible because of the pool of available savings; and the crisis of 1965 reflected the fact that this pool dried up, consumed by dizzying price rises.

These then were the structural features of the production system in the early 1960s which prevented supply adjusting to a demand which grew autonomously – and despite the impediment of the credit system. The resulting price rise thus generated a surplus profit, appropriated initially by development capitals, massively transferred to landed property as increased site values, hardly transferred at all to the building industry and in which the banks scarcely participated.

During the crisis of 1965–8 demand was on the ebb. It could not pick up until after the structural change of the credit system – of which the central feature was a mortgage market underwritten by public-sector financial institutions – introduced in 1966. The market was glutted and production collapsed. A fall in prices took place, slashing development profits which were also being squeezed by the costs of financing the unsold stocks of dwellings. During this period, when the overall surplus profit retained in the sector fell and then stabilised, there was a profound change in the system of production alongside a redistribution of the surpluses among the agents and a change in the way prices were determined. In various direct and indirect ways residential development was subordinated to the banks. As a result of the pre-emptive drain of surplus in favour of the financial sector, the rate of profit in development fell and, in turn, transfers of surplus to landed property were stabilised. From then on this new division of profits in favour of the banks could only be at the expense of the other agents whose profits were reduced to more-or-less normal levels. This redistribution brought about a rise in 'costs', and thus a tendency to price increases, which was quite independent of the state of demand.

This is effectively what happened in the rapid expansion of 1972-4 and also in the crisis which followed. Under the dual influence of growing short-term loans to developers and rises in the interest rate, financing charges grew considerably and cut into the developers' margins. The rise in prices was restrained by the credit system: borrowing costs were going up and intending purchasers, limited in their borrowing power, simply could not absorb major price increases. Furthermore the long crisis which started in 1975 was accompanied by continuing rises in price – in sharp contrast to the previous crisis. The low level reached by development profit could no longer permit the growth of financing costs to be absorbed by further cuts in development profit.[7]

Despite all this it was the level of development profit which remained the regulator of output from this system of development. Output fell drastically, and in such a sustained way, that we can truly report the abolition of that system of development which began in the late fifties and dominated capitalist housing in France for twenty years. Limited henceforth to special kinds of operations, especially infill developments in established towns where the profit was due to social segregation, private developers effectively abandoned the mass market to other forms of production.

Construction Prices

Prices for construction show a long-term tendency to grow relative to general price levels but to fall relative both to the total costs of development and to selling prices. These tendencies are not, however, steady: their evolution is dependent both on changes in the production system within the sector and on the relations between the agents.

From 1954 to 1974 the real growth of building prices[8] exemplifies a familiar phenomenon: productivity growth is slower in construction than in other parts of the economy, especially other production industries. But this has not prevented prices from falling or stabilising at times – due either to the occurrence of productivity gains which are then passed on (in part or in whole) as price reductions or, alternatively, to market conditions which oblige developers to accept falling prices in combination with rising costs and thus a squeeze on their profits. Thus, since 1950, a number of major periods become distinct:

(1) Up to 1955 a relatively rapid rise in building costs, linked both to rapid productivity growth elsewhere in the economy and to the the specific problems caused by the limited market for construction.
(2) From 1956 to 1962 prices stabilise as production expands – without much effect on prices because there were significant productivity gains.
(3) By contrast, between 1963 and 1965, building prices rose and by rather more than the main contributory elements. Building firms benefited from strong market conditions and probably increased their margins.
(4) Building prices stabilised again between 1966 and 1972, even falling detectably in 1966-7 and in 1969-70. This stabilisation was accompanied by rising prices for certain inputs – notably building materials.
(5) The years 1973 and 1974 were a time of rapid building price increases, but not sufficiently rapid to compensate firms for the even faster rise in materials prices.
(6) A new stabilisation took effect in 1975, associated with very unfavourable demand conditions facing firms.

If we now compare the selling prices of houses with the costs of construction we see that the growth of prices was significantly faster than the growth of building costs.

Right through the various conjunctural phases, and across regions and branches of construction, there is a strongly-marked tendency for building cost to form a shrinking proportion of total development cost (see Table 2.2), and to a lesser extent of selling prices (Table 2.3).

How then are we to explain this phenomenon when building prices are rising in real terms? The explanation is that the other components of the total development cost have risen even faster. But which ones: site acquisition costs or overhead expenses? This is what we shall now examine.

Table 2.2: Changes in the Structure of Development Expenditure: Paris, Paris Region and the Provinces, 1955–75

% of total development expenditure in the region

	Element [figures in brackets refer to definitions in Table 2.1]	Sector supported by loans from the *Crédit Foncier*												Unsupported sector									
		1955	1957 1962	1963	1961 1964	1965	1966	1964 1967	1969	1971	1971 1974	1970 1975	1975	1957 1962	1962 1964	1963	1965 1968	1969	1971	1973	1971 1974	1970 1975	1975
Paris Region	Developed site acquisition [15]	14.7		19.4	19		26.6	20			20.4	20.8									24.7	21.2	
	(of which: site price) [1]										(12.3)	(10.2)									(19.2)	(12.3)	
	Construction cost [16]	78.9		73.5	71		66.8	63			60.5	60.0									54.7	58.1	
	Overhead expenses [18]	6.4		7.1	10		6.6	17			19.1	19.2									20.6	20.7 •	
	(of which: financing costs [8]										(10.8)	(8.1)									(9.4)	(4.7)	
	Total development expenditure [19]	100		100	100		100	100			100	100									100	100	
Paris	Developed site acquisition [15]										25.7				25.4	22.3	25.2	24.6	23.4	24.9	26.6	22.2	24.9
	(of which: site price) [1]										(20.4)			(21.3)		(18.7)		(21.8)	(20.8)	(22.0)	(22.0)	(16.7)	(22.1)
	Construction cost [16]										57.3					66.8		20.9	59.3	54.5	52.1	56.5	50.1
	Overhead expenses [18]										17.0			7.7		10.9		14.5	17.3	20.6	21.3	21.3	25.0
	(of which: financing costs) [8]										(9.2)			(0.6)		(1.5)		(1.4)	(5.2)	(4.7)	(10.4)	(4.1)	(6.4)
	Total development expenditure [19]										100			100	100	100		100	100	100	100	100	100
Paris Region outside Paris	Developed site acquisition [15]								22.4	23.4			24.2			17.6		21.1		21.8			
	(of which: site price) [1]		(11.2)						(12.4)	(12.6)			(10.6)			(13.2)		(16.9)		(16.5)			
	Construction cost [16]								66.0	61.9			61.6			72.5		66.5		61.7			
	Overhead expenses [18]		7.2						11.6	14.7			14.2			9.9		12.3		16.5			
	(of which: financing costs [8]		(0.6)						(1.1)	(5.2)			(4.0)			(2.5)		(2.3)		(5.3)			
	Total development expenditure [19]		100						100	100			100			100		100		100			
Provinces	Developed site acquisition [15]					18.0		17.7	17.0							14.7							
	(of which: site price [1]					(7.8)		(10.4)	(11.1)							(10.1)							
	Construction cost [16]					76.5		71.2	69.2							76.3							
	Overhead expenses [18]					5.4		11.1	13.7							9.0							
	(of which: financing costs) [8]					(0.7)		(2.0)	(2.8)							(2.3)							
	Total development expenditure [19]					100		100	100							100							

Table 2.3: Changes in the Structure of Selling Prices and Relation of Surpluses to Capital Employed: Paris and Paris Region, 1957–75 (New private dwellings unaided by *Crédit Foncier*)

Element [figures in brackets refer to definitions in Table 2.1]		Paris							Paris Region outside Paris			Paris Region
		1957-1962	1965	1969	1971	1973	1975	1970-1975	1965	1969	1973	1970-1975
Structure of the selling price (% of selling price)	Developed site acquisition : [15]		17.7	19.6	20.0	21.8	22.6	19.4	15.1	18.2	19.2	18.9
	(of which: site price) [1]	(18.0)	(14.8)	(17.4)	(17.8)	(17.2)	(20.1)	(14.8)	(11.8)	(14.4)	(14.5)	(11.0)
	Construction cost [16]		52.9	48.7	50.7	47.7	44.3	49.3	62.1	57.4	54.4	52.0
	Overhead expenses [18]	6.5	8.6	11.6	14.8	18.0	22.7	18.6	8.5	10.6	14.5	18.6
	(of which: financing costs [8]	(0.5)	(1.2)	(1.1)	(4.4)	(4.1)	(5.8)	(3.6)	(2.1)	(2.0)	(4.7)	(4.2)
	Net development surplus [12]	15.5	20.8	20.1	14.5	12.5	9.2	12.7	14.3	13.8	11.9	10.5
	Total: selling price [20]	100	100	100	100	100	100	100	100	100	100	100
	Gross development surplus [21]	21.5	24.7	24.5	22.8	19.2	17.8	21.9	19.4	17.3	18.0	20.0
Finance (as % of total development expenditure [19])	Developers' capital		18.8	18.0	15.9	18.5	15.5		21.4	16.4	17.4	
	Banking capital		13.2	33.6	29.9	31.6	31.5		12.7	27.2	29.3	
	Total capital employed [*taux de couverture*]		32.0	51.8	45.8	50.0	47.0		34.1	43.6	46.7	
Gross development surplus [21] ÷ developers' capital X 100 [*taux de rentabilité*] ('rate of return'[a])			163.3	162.7	167.4	128.5	184.0		137.4	137.9	135.4	

Note: a. This indicator of return takes no account of time, i.e. of the speed of rotation of developers' capital.

Development Margins, Site Prices and Bank Interest

Site Acquisition Costs

The unbridled rise in site costs has often been blamed for the rise in house prices and the shrinking proportion of building costs within the total. From the mid-1960s to the early 1970s there were cries of alarm over this increase, which looked set to absorb a major part of national resources devoted to construction. One expert foresaw that land costs would rise from 25 per cent in 1970 to 57 per cent in 1985 (of the total development cost of a dwelling) (Derycke 1972: 99). Such projections were not challenged by the best development economists (cf. Granelle 1975: 68; Lipietz 1974).

More recent empirical enquiries do, however, question them. Granelle found that land costs as a proportion of house prices tended to stabilise after 1965[9] and this conclusion has been confirmed by all other studies. These data invite some deeper theoretical reflection about the changes in social relations which the data, in their own way, imply. First, however, I will summarise the main features of the data.

The relation of the price of developed sites to the total development cost per m^2 of dwelling is impossible to establish prior to 1965, the data becoming better and more comprehensive after that. It would be good tc examine these for each urban area but the data only permit Paris and its Region to be distinguished from the provinces.

In the Paris Region we see a marked growth in the price of developed sites as a proportion of total development cost up to 1965, and then a slowing down, indeed a stabilisation.

In Paris itself, the centre of the agglomeration, it seems that the proportion continued to grow even after 1965.

In the provincial towns the stabilisation of site costs relative to total costs after 1965 is very marked.

The prices of serviced sites are the sum of two main components, both related to the floorspace being built: on the one hand the pure land acquisition cost, on the other hand the cost of preparing for development (compensation for displacement, evictions, demolition, installation of infrastructure, certain local taxes and the costs of any land or works which have to be donated for public uses).

Practitioners and researchers agree that pure land price as a proportion of the price of serviced sites rose substantially between 1955 and 1965, though quantitative data are absent. It appears that after 1965 this element was more stable than the total price of serviced land. It seems that in the early 1970s the proportion was 21-22 per cent in Paris, and 11-13 per cent or 16-17 per cent in the rest of the Paris Region (the two ranges reflecting the difference between subsidised and unsubsidised sub-sectors respectively).

Examination of the makeup of site costs after 1965 shows in addition a tendency for land price to diminish relatively and for the other elements to go up. This is particularly clear in the sub-sector subject to subsidies and state loans. This undoubtedly reflects divergent location practices for 'subsidised' and 'unsubsidised' housing. The former tended to be built increasingly on peripheral sites, less well served; the latter remained in more central or accessible locations – or at least tended to gravitate back to such sites after the dispersal evident in the 1960s.[10]

It seems then that the increase in the price of serviced sites was greater than that of total development costs or of final selling prices until about 1965. In the following decade it grew at about the same rate as selling prices. In this later period the growth of the price of land itself (per unit of floorspace built) was slower than that of the clearance, servicing and associated works.

The Development Surplus and the Financing Cost

The data on these two remaining elements of the selling price are scanty and relatively recent. Here again the production of statistics is very much a social phenomenon. The developer's profit was very much a Loch Ness Monster,[11] sustaining a great burden of polemic during the property boom. No real studies, and then only very limited ones, were begun until the mid-1970s. Civil servants were preoccupied with the inflation in overhead costs and the professionals of the development industry were much more open with information on developers' margins when those were falling and on financing costs when those rose.

In fact it is this combined effect which the data (in Table 2.3) show us. Development surpluses in Paris, in the unsubsidised sector, seem to have grown from 21-22 per cent of selling price in 1960 to 25 per cent in 1965 when the crisis was triggered off. After 1965 there is a stability in the margins underlying some fluctuations: in 1975 the margin settled at 17-18 per cent of selling price. The data for the rest of the Paris Region are fragmentary and do not yield a clear picture.

While development surpluses tended to diminish, financing costs grew markedly. In the early 1960s they were negligible and they remained small through the decade. It seems that they began to account for a significant part of prices in the early 1970s: by 1971-4 they represented 9-11 per cent of selling prices in the Paris Region, irrespective of location or sub-sector.[12] The tendency to growth seems to have abated fairly soon, however, and was weak from 1965 to 1970, strong thereafter.

This growing element reflects a phenomenon which appeared on a massive scale from the late 1960s: the use of bank credit to finance developments. It began in about 1965 but was not widespread until about 1968. Reinforcing the effect of increased credit, the rate of interest on loans to developers rose from 9 per cent in 1965 to 12-15 per cent in 1975 (Pascal 1971: 258; CEGI 1977: 54) and the combined effect was a massive new cost for developers.

Resumé of Results: House Prices and their Components

We can draw all this evidence together in the form of a coherent periodisation.[13] Before 1965, and especially in the 1962-65 boom, the real growth of house prices was very rapid: developers oriented their production increasingly towards the luxury sub-markets while the first systems of bank lending for house purchase were getting established and while state funding was dwindling. The boom in sale prices and in output were accompanied, until 1962, by stable construction prices in real terms. They increased in 1963-5, but much less fast than house prices. The result of all this was the inflation of the real development surplus, reinforced by the sustained high level of sales and the fact that purchasers made advance deposits. The resulting high rate of profit in development brought a rapid rise in the price of sites – clearly more rapid than the growth of house prices themselves. Site costs thus came to account for a growing part of final prices – and the main ingredient was the actual price paid for land. A speculative mechanism resulted whereby the growth of development surpluses fed the bidding-up of site prices, which in turn led to increases in selling prices for dwellings as the developers tried to preserve their margins. This inflationary spiral came up against the limits of what the market could pay – limits which were narrowed by changes in the way house purchase was funded. A severe economic crisis broke in 1965, reflecting the structural collapse of that system of development which we can describe as essentially petit-bourgeois and based substantially on inherited wealth.

During the 1965-8 recession the selling prices of new houses fell in real terms, while the costs of building stabilised. Development surpluses declined and the financing of development projects began to change its form. Sales fell and the growth of unsold stocks led to the beginning of bank advances to developers. Rates of interest were moderate and financing costs remained limited. Falling rates of profit in development began to weigh on the prices of sites – and this element was thenceforth to be a stable component in the prices of privately-built dwellings. In this period a new system of relations between the agents becomes established: a system we can describe as dominated by financial and monopoly interests.

During the 1968-74 expansion, sale prices rose very slowly, and building costs remained stable, until 1972. Development surpluses never regained their pre-1965 level but clearly recovered to some degree and that remained the position until production took off again. It was only from 1973 that real price increases for new houses became rapid, though nothing like the boom of 1962-5. This new boom was accompanied neither by growing real development surpluses nor by relative growth of site prices: the proportion of site cost in final sale price remained broadly constant while the pure land element itself diminished relative to the costs of servicing. Here the rise in house prices is generated by two new forces. On the one hand construction prices grew rapidly in 1973-4, pushed up by the price of building materials: this constituted a transfer of profits from the building industry to the – much more highly monopolised – materials and components industries. On the other hand the financing costs weighed heavily on developers: the combined effect of their reliance on bank finance for projects, widespread since 1968, and of the rapid rise in interest rates from 1973. This is a sign of the massive transfer of development profit in favour of the banks. The period 1968-74 saw the expansion of the new, and essentially monopoly-dominated, system of development.

The crisis which began in 1974 can be seen as precisely the crisis of this new system. At the time of the 1965-8 crisis there was a dramatic fall in output. But this time selling prices continued to rise in real terms, albeit at a reduced rate. Building costs stabilise and development margins shrinking – to the point of bringing on the recession. Site costs remain constant but financing costs continue to grow because of the slow rate of selling. The drain which the banks have imposed on the development surplus seems to have brought it down to a minimum rate of profit; from this point onwards any growth in financing charges are a cost factor, adding to production costs and sale prices – even in the recession.

This interpretation of the movements of prices and of their components places the rate of development profit at the heart of the explanation. This concept permits the identification of the overall mass of profit available for appropriation by landed interests and by circulating capitals – strongly related to the volume of circulating capital involved. It is thus a concept fundamental to any empirical analysis of land rents.

Notes

Translation by Caroline John and Michael Edwards.

1. This section draws upon principles for the analysis of urban development and of urban rents which have been developed elsewhere. See Topalov (1981).

2. A critical review of studies up to 1975, and a remarkable interpretation of their results, are given in Ansidei, Carassus and Strobel (1976). In addition to the studies reviewed that I have used various scattered and less systematic data for the period up to 1975 and some studies published subsequently.

3. This is not an exhaustive list of the problems posed by these surveys. The data on sale prices are generally estimates rather than actual records of transactions. The expenses, which are spread over several years, are sometimes expressed in terms of prices ruling in the first year, and sometimes not capitalised but merely added.

4. It must be said that throughout the 1970s there were formal and informal interactions between those responsible for research at the heart of the French administration and empirical researchers in the Marxist tradition. I believe that this contact helped to define the content of the data I am using here. I shall draw heavily, for example, on the excellent work of Strobel (1980) undertaken within the administration but reflecting these interactions.

5. Private development in the early 1960s did not have the monopoly position in the Paris residential market which it was to achieve. later. Its activity began, however, to transform the Parisian market in such a way as to transform monopoly rents deriving from segregation and privileged neighbourhoods. This point is essential to an understanding of the later rapid price rises in Paris, in 1973–6 and since 1980.

6. By 1968, six years after the Evian agreement, 20.4 per cent of the returning households owned their own homes in France: 54,500 of them.

7. The growth of bank lending to developers while unsold stocks were mounting had the effect of creating a floor below which prices could not fall. The slump led to increased financing costs and to price rises. The banks' dominance grew with the depression which they were fuelling. Their method of taking property as securities, furthermore, allowed them to control prices directly. As one developer said in 1976 'The real owner of my stock is my bank' (CEGI 1977: 63).

8. I have used the only price index which is available over the whole period: the INSEE index for *'travaux traités'* deflected by the PIB as a measure of general inflation.

9. 'There is a change in the rhythm of growth in land prices if we compare the period 1965–75 to the accelerating growth before then . . . It seems that the development crisis of 1965 constituted the first turning point for land prices since 1950' (Granelle 1976: 51–2).

10. This evolution of locations is observable in the Paris Region during the crisis of 1965–8. See Topalov (1974).

11. The phrase '*ce serpent de mer*' was used in the late 1960s by Jean-Claude Aaron, President of the *Fédération Nationale des Constructeurs Promoteurs*.

12. I am concentrating on the estimates of the DBTPC (1971-4), the only research based on the accounts for completed projects. The estimates by CNEIL (1965-75) are based on *ex ante* budgets and could seriously underestimate these charges.

13. A comparable analysis, concentrating on the periods of inflation, has been conducted by Ansidei, Carassus and Strobel (1978: 151-8).

References

Ansidei, M., J. Carassus and P. Strobel (1976) *Bilan des enquêtes et études existantes sur la structure du prix du logement*, Ministère d'Equipement, Service des Affaires Economiques et Internationales, no. 26, Paris, mimeograph

—— (1978) *Logement: pourquoi la hausse des prix? Evolution 1960-76*, La Documentation Francaise 169, Paris

CEGI (Compagnie d'Etudes Economiques et de Gestion Industrielle) (1977) *Enquête auprés des professionels*, Paris

Combes, D. and F. Imbert (1978) *Industrie du bâtiment et immobilier: la production de logements des grandes entreprises et leur clientèle*, Centre de Sociologie Urbaine, Paris

Derycke (1972) Supplement to *Géomètre*, no. 7, July, 73-124

Dhuys, J.-F. (1971) 'Etude économique simplifiée d'une opération immobilière de charactériques moyennes à Paris' in *La vie urbaine, 4*, 263-271

—— (1975) *Les promoteurs*, Seuil, Paris

Granelle, J.-J. (1975) *Le valeur du sol urbain et la propriété foncière: le marché des terrains à Paris*, Mouton, Paris and The Hague

—— (1976) *Evolution du prix des terrains à bâtir de 1965 à 1975 dans quelques agglomérations françaises*, Fédération Nationale des Promoteurs Constructeurs, Paris

Harvey, D. (1982) *The Limits to Capital*, Blackwell, Oxford

Lafont, J. and D. Leborgne (1977) *Immobilier et processus inflationnistes*, CEPREMAP, Paris

Lipietz, A. (1978) *Le tribut foncier urbain*, Maspero, Paris

Massey, D. and A. Catalano (1978) *Capital and Land: Landownership by Capital in Great Britain*, Arnold, London

Pascal, F. (1971) *Economie de la production de logements*, Université Paris I, Paris

Pottier, C. and M. Rogalski (1977) *Financement et coûts de la promotion immobilière privée*, Centre d'Etude des Techniques Economiques Modernes, Paris

Secchi, B. (1971) 'Il settore edilizio e fondario in un processo di sviluppo economico' in F. Indovina (ed.), *Lo spreco edilizio*, Marsilio, Padova, 3-46

Strobel, P. (1980) 'Coûts, financement et stratégie de la promotion privée dans les années 1970' in *Les Cahiers du Ministère de l'Environment et du Cadre de Vie, 1*, 20-30

Topalov, C. (1974) *Les promoteurs immobiliers: contribution à l'analyse de la production capitaliste du logement en France*, Mouton, Paris and The Hague

—— (1981) *Le profit, la rente et la ville*, Centre de Sociologie Urbaine, Paris

3 HOUSING DEVELOPMENT PROCESSES IN EUROPE: SOME HYPOTHESES FROM A COMPARATIVE ANALYSIS[1]

Marino Folin

Post-war Trends in Housing Output

Since the end of the Second World War more dwellings have been built in Europe than ever before. To quote the examples only of the countries I shall be referring to later, the period between 1945 and 1982 has seen the completion, in France, of 11.6 million dwellings, equal to 48.9 per cent of the 1982 housing stock; in Great Britain, of 10.5 million, equal to 49.3 per cent; in West Germany, of 17.5 million, equal to 67 per cent; in the Netherlands, of 3.4 million, equal to 67 per cent; in Sweden, of 2.6 million, equal to 69.3 per cent.

In rather less than forty years the housing stock of every country in Europe has been extended enormously – part of the stock existing at the end of the war has been replaced and part modernised. As a result of this unprecedented production, housing conditions in all these countries have, in general, improved considerably from the point of view of overcrowding, the provision of services and physical standards. The housing stock of almost every country is now, in aggregate terms, greater than the number of families, or (in the cases of the Netherlands and West Germany) only slightly less.

This enormous quantitative increase in production has not taken place by chance. It reflects accurately the way the housing problem has been seen in all the countries of Europe ever since the immediate post-war years: as a quantitative problem – the absolute lack of enough housing. Housing policies have been, and explicitly so until the early 1970s, basically quantitative policies. The solution to the housing problem, defined in terms of absolute shortage, has everywhere been seen as lying in a greater number of new dwellings being finished each year (UN 1980; Donnison and Ungerson 1982). Housing policies in the various countries have long been comparable because they may all be summed up in terms of a 'dwelling target'. The centrality of this target can, for example, be seen in the way in which housing problems have been presented in the reports for the United Nations prepared, annually at first and then periodically, by the Economic Commission for Europe from 1953 onwards (UN 1954–1963). These reports compare the extent of the shortage in the various countries and pay particular attention to the means by which shortages could be reduced and the potential barriers to such reductions. Emphasis is consequently placed on the number of dwellings produced annually, and on the costs, ways and means which have permitted or prevented the highest possible production levels.

Before going any further we must consider briefly two aspects of this quantitative production in order to avoid confusion.

The first concerns the nature of the housing produced. To state that housing policy and production in Europe have long been dominated by the problem of quantity does not mean that there have not also been redistributive and qualitative targets. On the contrary, this quantitative production has not been aimed at the population in general but has given precedence to the interests of certain social strata rather than others. Clearly, it has not been aimed at the interests of classes as defined by Marx with reference to production relationships, but has given precedence to the interests of certain social strata defined with reference to income levels or to particular social categories as defined by Weber.[2] The 'quantitative' policy, therefore, was not a 'working-class' policy (just as it was not a 'ruling class' policy either), but rather a 'social' policy that was an amalgam of a complex set of political forces, encapsulated most clearly in the programmes of the European social democratic parties in the post-war era. In operating thus, the houses produced under such policies have certainly had distributive effects. Housing policy, in other words, was also a 'redistributive' policy.

The second aspect concerns the reason for the high levels of production. There is nothing automatic, nothing mechanically inevitable in the fact that a shortage of houses should of itself lead to a quantitative policy and high levels of production;[3] just as there was nothing natural, nothing objective in the fact that there would be a shortage of houses in the first place. What is more important is to analyse the historical conditions under which housing shortages are 'seen' to exist, and in comparison with times when similar objective circumstances do not create such a subjectively perceived 'shortage'. Support for this point of view emerges from an examination of how the housing shortage was seen in the first United Nations report on the housing problem in Europe at the end of the War (UN 1949). This report argued that the shortage was caused only in small part by military action during the war, and it did not consider spatial variations in the availability of housing. For the most part, therefore, the housing shortage emerges from and is measurable against subjective definitions of housing standards.[4]

To define the housing problem in terms of shortage is therefore to make a political choice. The reasons for this choice were summarised by G. Myrdal in his Foreword to the UN Report of 1949. 'Housing', says Myrdal, 'is more than a grave social problem; [it is] a strictly economic problem'. That houses should be built is essential for the economy of a country for two reasons: in the first place, the level of economic activity 'depends in large measure on the availability of housing accommodation for workers who must man essential industries and agriculture', and, secondly, because 'building industry can play a considerable part in mitigating those economic depressions . . . which all governments are now pledged to try and avoid' (UN 1949: IV). Behind the production of housing there is therefore the problem of the reproduction of the workforce and the problem of economic recovery through the multiple effects on earnings and, therefore, aggregate demand which, in Keynesian terms, one would assume housebuilding investments have, just like any other investment.[5]

The emphasis on quantitative production is not the only factor common to many West European countries. Long-term trends in new housebuilding are also similar, although, of course, there are relative inequalities in terms of annual completions due to short-term cyclical dissimilarities. If we compare graphs of the annual production of dwellings (Figure 3.1) all the countries considered have a similar output periodicity. In the first period following the Second World War housing production increased rapidly. In Sweden, this period started earlier than in other countries because the rhythm of housebuilding was maintained there throughout the War (Wendt 1963); West Germany and France started a little later and West Germany (though not France) quickly made up for lost time. A second period, sub-divided into short cycles which vary from country to country, covering the 1950s and a part of the 1960s, features a slowing down of the rate of increase of housebuilding, though output was maintained at a high level. After 1957 the UN reports cited earlier note that production seemed to have reached a production capacity ceiling, at least in some countries (UN 1958: 39). In the third phase, after a short period of decline which

Figure 3.1: Dwellings Completed: West Germany, France, Great Britain, Netherlands and Sweden, 1946-82

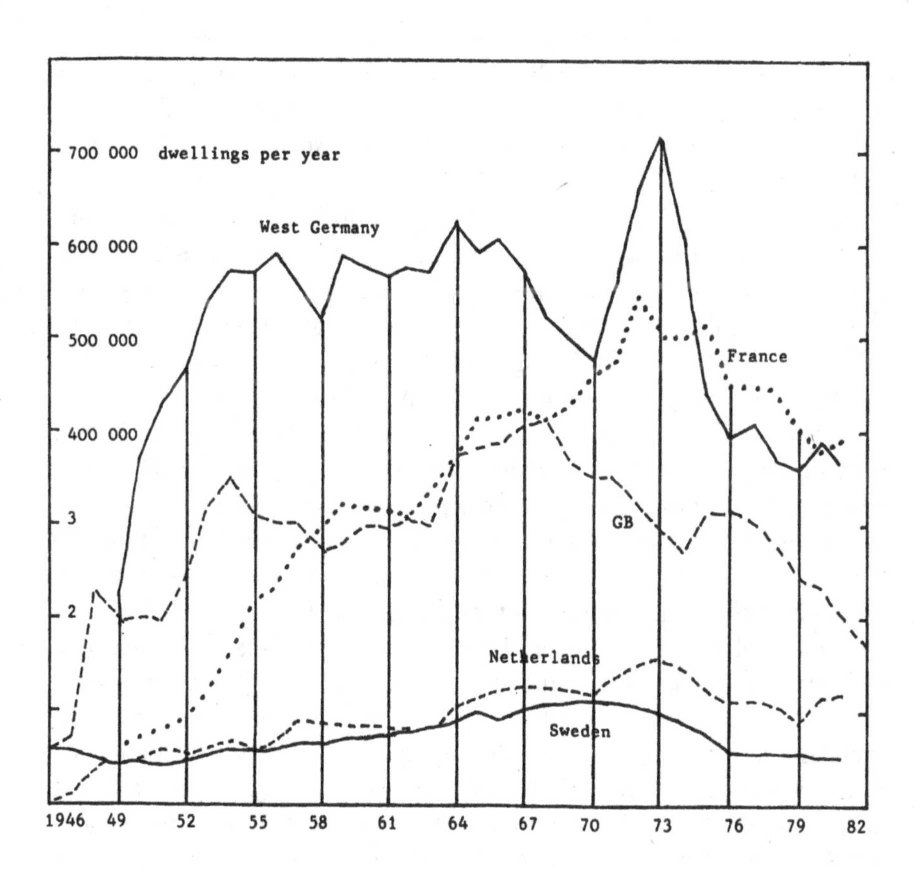

appeared at various points during the second half of the 1960s, production suddenly revived in all countries and reached hitherto untouched levels. In France the high point came in 1972, in West Germany and the Netherlands in 1973, in Great Britain in 1968 and in Sweden in 1969. After these peaks a fourth phase began in the second half of the 1970s in which housebuilding declined steeply in all countries, falling to the levels of the 1950s for most countries (1960s' levels in the case of the Netherlands, and the 1920s' levels in the case of Britain, if the war years and their immediate aftermath are excluded).

Problems with Orthodox Explanations of the Decline in Housing Output

The explanation generally given in government circles for the trends described above are perfectly consistent with the way the housing problem was originally formulated. Thus, on the one hand, the collapse in output levels in the fourth phase since the mid-1970s is considered, at least in part, as the inevitable result of the fact that there is no longer an absolute shortage of houses. Evidence for this claim is adduced in the (sometimes high) number of unoccupied dwellings and from the dwelling/household ratio, which is above unity. On the other hand, the fall in housing output is blamed on the economic crisis which Europe went through in the 1970s and the early 1980s, and on the deflationary policies pursued during the period.

Neither of these explanations is satisfactory. The first, related to housing needs, does not take sufficient account of the fact that, even when there are more dwellings than families, and even when there are unoccupied dwellings, serious unsatisfied housing needs continue to exist. Neither does it take into account the fact that, parallel to the improvements in the housing situation on a national scale, there is a worsening of the situation on a local and regional scale. For growing numbers of households, finding housing accommodation or simply maintaining possession of what they already have is a problem; thus there is a decline in housing conditions, even by reference to traditional housing standards, whenever they are applied locally and not just nationally. Housing problems appear at a national level too when proper weight is given to the changed housing needs consequent upon the general processes of economic crisis and restructuring which took place during the 1970s and 1980s. The second explanation, related to macroeconomic factors, ignores the fact that, even in those countries where some of the macroeconomic conditions present before the crisis have been reconstituted, such as a reduced inflation rate or a revival of the growth rate in production and investments, this has by no means been accompanied by a concomitant revival of investment in new housing, except for temporary cyclical upturns, as occurred in Britain during 1983.

There are cycles in housing policies similar to those in housing production. In particular, from the end of the 1960s onwards, there was a clear change in all the countries of Europe from policies supporting the production of housing towards policies supporting the demand for housing. This change has also been described as a shift from 'quantitative' policies to 'distributive' and 'qualitative' policies (UN 1980; Donnison and Ungerson 1982). The words used to describe this change are the same in all the countries: in France, *de l'aide à la pierre a l'aide à la personne*; in West Germany and in the Netherlands from *Objektforderung* to *Subjektforderung* and so on. Independently of the specific techniques used, what this change means is the same in all the countries: the 'liberalisation' of prices and rents (in both public and private sectors); cutbacks in financial aid and investments in the private or social housing sector; an increase in individual benefits, directly (via subsidies) or indirectly (via tax reliefs), in order to stabilise family spending on accommodation (whether rented or owned). The political significance of this change continues to be the subject of debate and research.[6]

The passage from a 'quantitative' to a 'redistributive' policy is part of the more general processes described above, whereas the decline in housing production is also due to the decline in public housing. Yet this change was not a sudden one: it began very early and some of its aspects, as I have argued elsewhere, were implicit in the social policies of the 1950s (Folin 1981) and were already noted since 1954 by the UN reports (UN 1956: 18). The decrease in housing production from the mid-1970s on can therefore not be explained solely by recent policy decisions. Nevertheless, amongst the political forces associated with the governments of most of these countries, the suggestion that there should be a radical liberalisation of the housebuilding sector has gained ground in recent years. It was envisaged that the abolition of restrictions and special conditions of the town-planning, tax and credit kinds would be able to guarantee the stability, if not the recovery, of the sector and, by means of filtering processes, solve the housing problem. According to this theory only residual traces of the problem would remain and these could be treated with *ad hoc* welfare-type policies. With every day that passes this theory is proving more and more misleading. In fact, the present period seems to be characterised, as far as housing is concerned, by the impractical nature of the policy decisions being taken, or rather their failure, since the results being produced are often the very opposite of what was hoped for (Harloe and Martens 1984).

The Economic Characteristics of Housing Production

For alternative explanations of trends in housing production, of their significance and the possible effects on them of policy decisions – whether specifically of housing policy or of more general economic policy – it is necessary to examine the sphere of production itself under separate headings according to the housing promoter in question. But before proceeding we must consider the special nature of house-building investment and the meaning of development.

In the works of economic theorists in the field of economic policy and in national accounts tables there would seem to be no doubts: housing is an investment, in the sense that the production of housing, like the production of machinery and means of transport, contributes to gross fixed capital formation (Harvey 1981: 69). It is the assumed equivalence between new housing output and other investment goods that forms the basis for the rationale of strategies or tactics of economic policy which use housing production as a potential macro-economic instrument. Such a view can have only a short-term validity however, for over a longer period, 'housing' and 'machinery' can no longer be equated in terms of being equivalent investments. Machinery, in combination with labour and under the command of capital, contributes to the production of 'added value'; housing, once completed, makes no such contribution, and still less so to surplus value. Houses lead, neither necessarily nor directly, to an increase in the productive capacity of a country.

Attention has already been drawn to this special nature of housing production by K. Wicksell in 1893, with reference to 'goods of greater durability (such as streets, railways, buildings, etc.)', which 'cannot be regarded or treated as capital in the narrower sense, but once they are there must be placed, economically speaking, *in the same category as landed property itself*' (my emphasis) (Wicksell 1954: 119). Housing, therefore, once built is like land. And, like land, its increase in price over time and its existence as a sphere of investment are results of the exercise of a monopoly and not of a combining with labour in a production process under the command of capital.

This peculiar economic characteristic of housing, which is independent of the nature of the activity undertaken in it or achieved by means of it, is a quality which was observed very early. In his *Apology for the Builder* (1685), N. Barbon wrote against the policies restricting the growth of London and identified the origin of house values in the growth of the city: 'New Buildings are advantageous to a City, for they raise the Rents of the old Houses. For the bigger a Town is, the more of value are the Houses in it . . . when a Town happens to be increased by addition of New Buildings . . . the old Houses . . . increase in value'.[7]

It is therefore the process of growth of the city which produces value; it is in the process of growth that houses become an investment.

What Brabon describes is, however, only one aspect of the process of increasing property values: the above-mentioned work refers to the value increase in houses already existing. What is the relationship between this process and the production of new houses? To provide an answer, we must briefly consider the meaning and nature of the development process.

In British Town and Country Planning Acts since 1947, the development process is described as 'the carrying out of building, engineering or mining operations in, on, over or under land, or the making of any material change in the use of any building or other land'.[8] From the physical point of view, 'to develop' is therefore used to denote a production process the result of which is the built environment. With the words 'to develop' is meant 'to produce the built environment' (Lichfield and Darin-Drabkin 1980: 65). Another characteristic of the physical aspect of this production is the fact that it takes place as 'the material change of any building or other land'. Again, 'land' and 'building' are equated. The object of the development process is the conversion of land use: when the Town and Country Planning Act, 1971, speaks of the development process it means 'land development'.[9] But why are processes of 'material change', or 'addition of new buildings' described as development? How do building and development relate?

In economics, the development process is described through the main functions accomplished in it: '[a developer] assembles the site . . . arranges finance . . ., gets the scheme built' (Harvey 1981: 70-3). As for the logic which the developer adopts, a distinction should be made between public and private developers: the former are 'non-profit-making' while the latter are 'essentially profit-motivated' (Balchin and Kieve 1979: 82). From this neoclassical economics point of view, the developer's profit is 'the return for the entrepreneurial risk-taking function' (Balchin and Kieve 1977: 87); the 'commercial' developer is 'an entrepreneur' (Bowley 1966: 329).

Such definitions are unsatisfactory. In particular, they fail to explain the origins of the profit generated, and they also fail to note the aspects which are common to the two forms of development; public and private. The Marxist critique of political economy enables the origin to be seen in the production process taken as a process of increasing value. The development process is not simply 'to produce built environment', but a peculiar form that this production takes historically, and is precisely production for surplus value. If we adopt this point of view, the development process appears as a process of increasing value, the dynamics of which may be understood only within the overall process of social production. The value present at the end of the development process, a part of which is appropriated by the developer as his/her profit, is partly produced during the process of development and partly taken away from other sectors during the same process. This process of production and redistribution of value is the result of the combined action of the forms of control over labour exercised by capital, and the forms of ownership of land and capital. These relations of production and distribution are contradictory and frequently in conflict with each other, whilst relations of distribution are subordinate[10] to the relations of production. Any examination of housing production in terms of type of promoter,[11] therefore, involves examining different development processes which combine elements of value increase (via production) and redistribution (via, amongst others, rents and interest charges).

Forms of Housing Provision

There are several typologies which recur more frequently than others amongst the forms of private and public housebuilding promotion occurring in Europe since 1945. Some of them are common to several countries, albeit with different emphases and working in different ways. Official statistics do not always make it possible to identify them sufficiently clearly and the same definitions given for each one do not always denote promotional forms which are exactly the same. I shall attempt below to bring together the different forms of promotion into groups which are as homogeneous as possible, but warn the reader that this is a first attempt at such a subdivision and that some degree of approximation is inevitable.

(a) Private Promoters. In Great Britain these are all classified under the 'Private Sector'; in West Germany 'private Haushalte' and 'sonstige Wohnungsunternehmen'; in the Netherlands 'partikulieren'; in France 'societés et particuliers'; and in Sweden 'Amdra byggherrar' (other investors). Yet, whatever name is adopted, this sector contains a plurality of forms of production which official statistics fail to distinguish satisfactorily. To achieve this we need to consult more sources of information and undertake sample and geographically localised surveys; it is thus very difficult to reconstruct temporal series which are complete and mutually comparable.

Within this sector we must distinguish at least the following sub-groupings:

(a.1) Individuals building their own houses. These are householders or individuals who start the process of development to produce a house for personal use and not for the market. In this case it is the ultimate user who arranges the various phases of the production process: purchasing the site, organising the financing and getting the scheme built.

Many different combinations are possible and consequently many agents may be involved with many different relationships between them. The land plot may already be owned by the ultimate user, or purchased for this purpose to build on. Finance may come directly from family savings accumulated over the years, or derived from the sale of inherited property, or from redundancy payments. Alternatively, it may come, wholly or in part, from a financial institution, either as a short-term credit or longer-term mortgage. The construction of the dwelling itself may be undertaken directly by the user, by labourers working directly under the orders of the user, by a private builder hired to build the house; or the house may be bought already built, either wholly or to the extent that it just has to be assembled on site. Depending on the case in question, the 'consumer-developer' identifies with one or more of the agents involved and is placed in conflict with the others.

This form of promotion is present to a significant extent in France where, in various forms, it has been more or less important in different periods (Topalov 1982), and to some extent in West Germany. The limits of this form are not always clear as they tend to blend into those of the second grouping in cases where the house is built for owner-occupation but, in part, is also rented out. This latter aspect may determine from the outset the typology and dimensions of the dwelling, as happens in West Germany where two-family houses are often built, one part of which is then let.

(a.2) Commercial developer, usually defined as the 'typical private-sector developer'. Here the developer starts a process of development that leads to the production of houses for the general market, either for sale or to let. In this case too there are numerous possible forms though the following situations are most common. The developer purchases the land, in large quantities in advance, or in small quantities as it is needed, and if necessary arranges for infrastructure work to be done on the site. The short-term capital market is resorted to for the necessary finance.

A construction firm is engaged to build the dwellings, and the developer arranges directly for them to be placed on the market. In Great Britain, there is the important figure of the builder-developer (the speculative housebuilder) who produces houses for the owner-occupied housing market. For assessment of their operations it is essential to distinguish their relative size (Harloe, Issacharoff and Minns 1974; Drewett 1973; Pahl and Craven 1975; Ball 1983). In France, on the other hand, the builder and developer are separate, and there is a tendency for the latter to be identified with the financial institutions (Topalov 1982). Besides having both the builder-developer and the pure developer, West Germany, the Netherlands and Sweden also have financial institutions which operate as developers – including insurance companies, pension funds, banks, etc. In such cases the dwellings tend more often to remain the property of the developer and are rented out.

(b) 'Non-Profit-Making' Housing Organisations. Again, such institutions take a number of forms. In Great Britain they are called 'Housing Associations'; in West Germany 'Gemeinnützige Wohnungsunternehmen'; in the Netherlands 'Woningbouwverenigingen'; in Sweden 'Kooperative Foretag'. In contrast to the typical private developer, these companies produce dwellings which remain their property and are rented out.

Variations in the characteristics of non-profit-making institutions arise because of their different origins and are described in the statutes of each. In Great Britain, for example, they are descended from the philanthropic associations of the second half of the nineteenth century; whereas in West Germany, Sweden and the Netherlands, their origins lie in the co-operative and trades union movements. There is consequently a variety of possible combinations, especially as regards relations with construction and financing agencies. Integration of the various phases is most highly developed in Sweden, where some of the largest co-operative companies in existence, such as SR, purchase land, furnish it with infrastructure facilities as necessary, and see directly to the financing and building of the dwellings (Wendt 1963: 88).

(c) Semi-public Bodies. They include, in particular, France's 'Habitations à Loyer Modéré' (HLM). In Sweden, the Netherlands and West Germany similar agencies, because of their institutional statutes, tend to fall within the category (b). Such bodies are semi-public because they arise from public initiatives. They exhibit many similarities to the

previous category. Generally they produce in order to let, although the HLM also produces for sale. Their activities are strictly state-controlled, especially as regards selling prices or rents, generally fixed in relation to a dwelling's 'production cost'. They enjoy special conditions for obtaining credit and receive subsidies from the state or local government. The HLMs often purchase land which has already been provided with infrastructure at the expense of the local authority. They have no capital of their own and must therefore negotiate directly (or through the local authority) with landowners, financial institutions and construction firms.

(d) Local Authorities, Municipalities, Gemeenten, Kommunen. Local authorities themselves operate as developers. As in the two previous cases they produce dwellings which in general remain their property – at least as far as initial intentions are concerned – and are rented out. As regards access to land, they can adopt a policy similar to that of the typical developer although they have rights of compulsory acquisition which private agencies do not possess. They tend to buy up large quantities of land and gradually set up infrastructures as necessary, although in recent years piecemeal infill development has grown in importance. Functions, which historically have been those of the separate landowners and developers, are here united in the figure of the local authority. In contrast to the practice with semi-public bodies and non-profit-making companies, the cost of the land and of the installation of infrastructures is wholly the responsibility of the local authorities themselves. Finally, they too may enjoy special conditions of access to credit and they receive state subsidies for housebuilding.

They have a contractual type of relationship with construction companies, and direct labour building activity takes place only to a limited extent (though once it was more common in Great Britain).

This form of public promotion is widespread, though more so in Great Britain (Merrett 1979) than in other countries where there are relatively greater numbers of non-profit-making or semi-public companies.

(e) Government Departments (Öffentliche Bauherren): Unternehmen ohne Wohnungsunternehmen. This last form mirrors the first: in this case we have public bodies or private enterprises which do not produce for the market but to house their own employees or particular categories of workers. The dwellings remain the property of the promoter and are rented out. They, like private entrepreneurs, generally buy land

which is ready to build on; in the case of large-scale operations they act rather like local authorities. Finance is obtained in whatever way the Department usually obtains finance, and housebuilding promotion remains a secondary aspect of its overall activity.

Having completed our classification of different types of promoter, some comments and qualifications need to be made at this stage. Group (a.1) corresponds to the system of production-circulation of housing accommodation which Topalov calls 'sans promoteur'; group (a.2) to what he calls 'promotion privée'; all the others, except for the last, to his 'promotion publique' (Topalov 1982). Even this general attempt at classification leaves some problems unresolved; where should we place the consumer-developer who lets part of his/her house? Where should we place some of the non-profit-making associations? The important thing to remember in each case is that the different forms of development should be grouped together in such a way as to reflect different processes of value-increase, that is of the production and distribution of value. How this value is formed, how it originates, may be reconstructed from an examination of how the different developers function in the context of the social and economic relationships within which they operate. This involves examining, with respect to the valorisation process of social capital, the forms of finance, their terms and their real and monetary cost; the technical features and productivity in the building sector (compared with the average) and the forms of labour control exercised therein; the nature of ownership and access to land; the level and types of public spending (including tax relief, subsidies, rebates, allowances, etc.); the structure of and temporal trends in rents and the level of demand in different housing markets; and, finally, the relations between the various forms of ownership and control in terms of their combination, identification, conflict.

As regards the role of land in the development process, relevant information is fragmentary. Some reliable data have been gathered, however, and it is on the basis of these that the following comments are summarised (UN 1973). (i) The cost of land in the post-war period in all the countries under consideration has risen on average more quickly than either the cost of living or of building costs;[12] (ii) the most rapid rises in land prices have coincided with active phases of development. The approval, for instance, of a town plan or other means of town planning seems to cause land prices to rise,[13] while, in general, the rate of increase has been greatest in suburban areas;[14] (iii) in almost all countries the share of land in housing costs has increased;[15] (iv) attempts to control the cost of building land (tax on land values, advance purchase of vast areas by local authorities) have been difficult to apply or have been shortlived; only in recent years, after 1975, have containment and control measures been approved to any great extent, and this only in certain countries (Sweden, UK and the Netherlands) (Massey and Catalano 1978; Headey 1978); (v) some differences in the relative rates in land prices should be noted: the Netherlands and Sweden have the lowest percentage increase, France, Italy and Spain the highest, while West Germany, Switzerland and Great Britain have percentage increases between the two extremes.

In an attempt to establish the causes of the phenomena described above and the different patterns which have emerged the ECE reports for the United Nations have tried to identify the correlation between the percentage increases in price observed and several variables such as the rate of per capita income increase, the cost of living index, population density and its growth rate, and tourist movements (UN 1973: 99-103). The results are mutually contradictory and disappointing. This was perhaps inevitable (and to some extent foreseen by the writers of the report), considering that the available data refer to national averages and cannot therefore take account of the very marked regional and local variations within any given country. It is, however, not only a question of a shortage of available data, but also of limitations which are inherent in the method adopted. Even where the existence of a marked correlation between two groups of phenomena is in fact identified, we still had no information as to its cause.

To achieve this, meaningful results might emerge from a comparative analysis which established how land prices developed in relation to the various forms of housebuilding promotion, starting from the hypothesis that the formation of land value is an integral part of the development process, not something which precedes it, which originates outside it, like a cost factor. Where analyses of this kind have been undertaken they have demonstrated that this hypothesis is not without foundation (Topalov 1982).

Historical Changes in Forms of Housing Provision

The various forms of the development process summarised above do not follow a constant trend in time, nor are they present everywhere to the same degree of intensity as regards the number of new dwellings completed annually.

In post-war Britain there have been two main forms of promotion:

Figure 3.2: Great Britain: Dwellings Completed by Form of Provision 1946-82

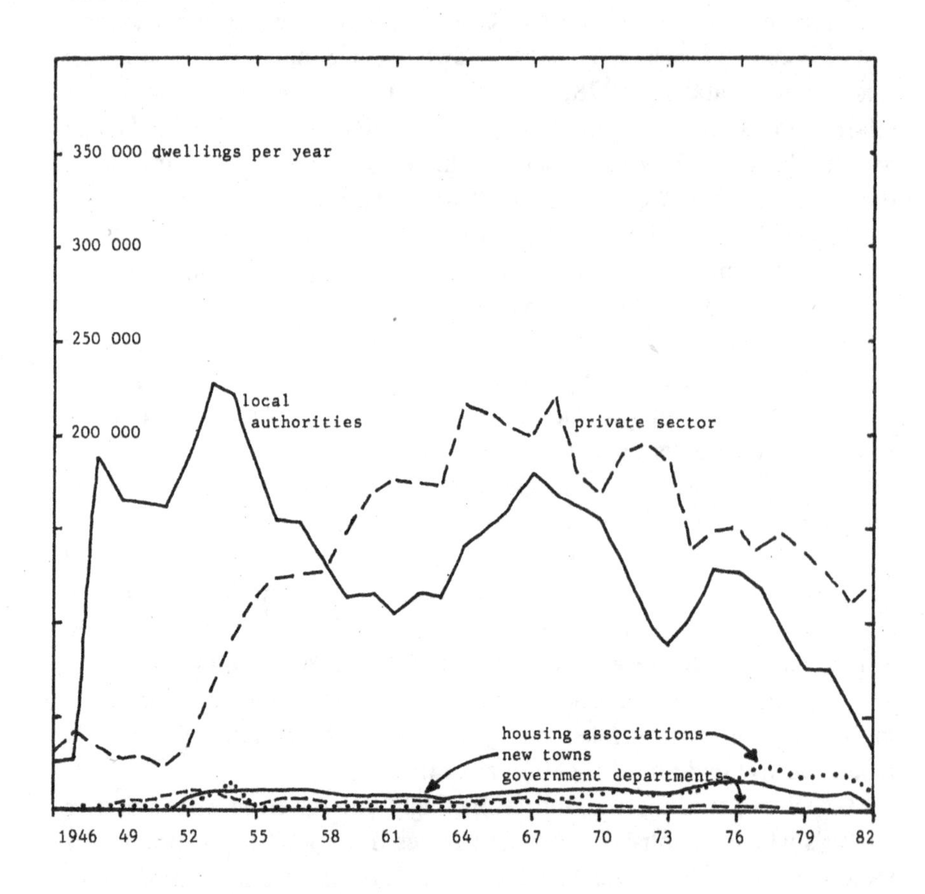

by the local authorities (group d) of the above classification – and by the builder-developer (group (a.2)). The former produce housing for letting and the latter for owner-occupation. Other forms of promotion such as Housing Associations (group (b)), Government Departments (group (e)) and New Town Corporations have made smaller percentage contributions to the total production (see Figure 3.2).

After the high point of 1953, the number of dwellings completed annually by local authorities tends to fall over time, with three short, regular and complete cycles of decreasing size with relative high-points in 1967 and 1975; a low-point was touched in 1981, and this represents an absolute low since 1947. The private sector features a single large cycle, with an absolute maximum point reached in 1968; the decreasing phase features numerous short, irregular cycles of variable size. If we

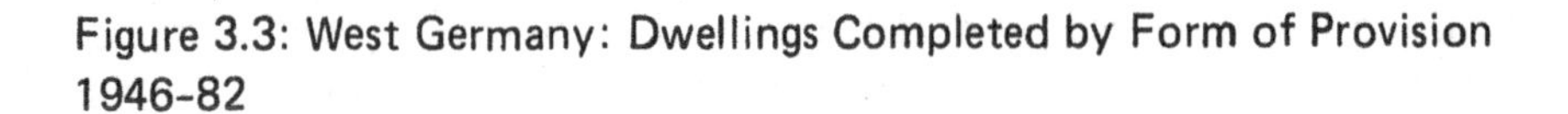

Figure 3.3: West Germany: Dwellings Completed by Form of Provision 1946-82

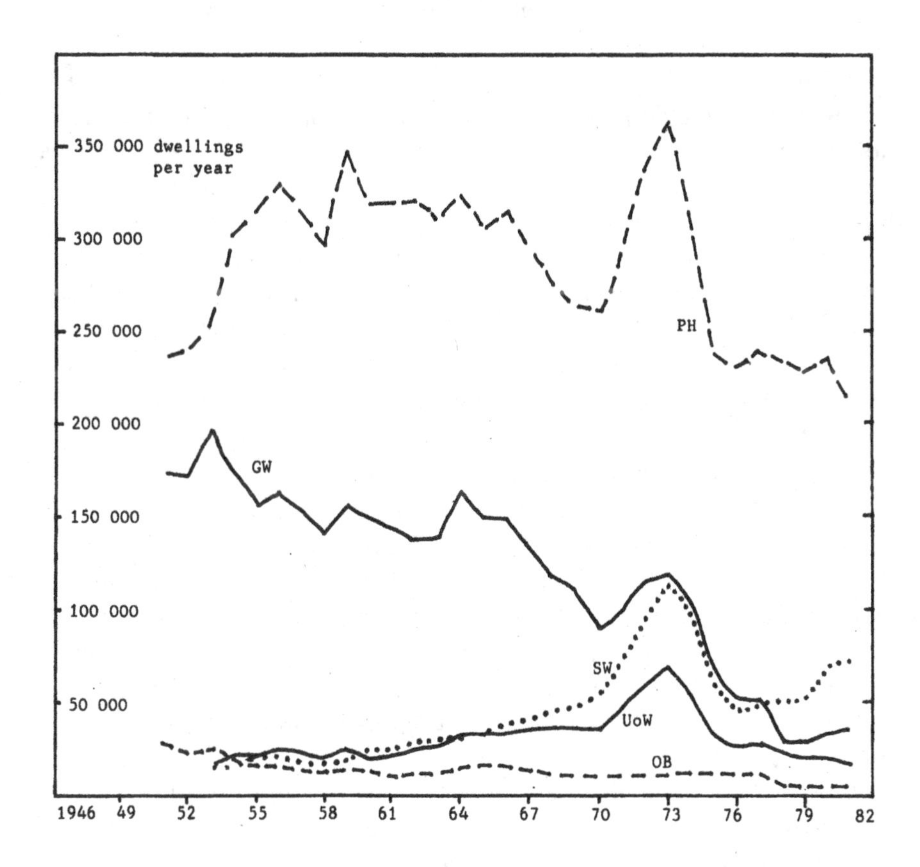

consider the two cycles together, they present opposite tendencies during the decade 1953-63, with the public sector on a constant downward trend and the private sector consistently rising; later, during the 1970s, the trend with both was to decrease, though there were short cycles of a different nature in each sector, staggered irregularly in time.

In West Germany there are at least four major forms of promotion, though they do account for a variable proportion of the total. They are represented by: *Private Haushalte* (PH) (a.1) and (a.2); *Sonstige Wohnungsunternehmen* (SW) and *Unternehmen ohne Wohnungsunternehmen* (UoW) (a.2); *Gemeinnützige Wohnungsunterenehmen* (GW). The first, PH, in its turn is made up of various forms of promotion which do not separately appear in the official statistics: consumer-developers or those who promote the construction of dwellings for their own use; private individuals who produce two dwellings, one of which is rented out; private individuals who invest in the production of houses for letting. The second, SW, is made up of builder-developers who produce dwellings for sale or letting on the open market. The third, Uow, is represented by building societies and insurance companies who produce houses which are generally for letting. The fourth, GW, is made up of semi-public companies, co-operatives or others, which produce dwellings for letting (see Figure 3.3).

Trends in all four sectors are quite regular and consistent from the mid-fifties on, with a single important irregularity, common to all four sectors, occurring between 1971 and 1974. The PH sector is the dominant one with a steep rise up to 1955, holding firm until the mid-sixties, and going into a steady decline after 1966, interrupted only by the 1971–4 boom. The GW sector was for a long time the second most important housebuilding promotion sector. Quite high levels of activity were maintained until 1965, after which there was a decline parallel to the decline in the private sector until 1975, followed by an even steeper decline thereafter. The other two sectors, SW and UoW, operated much less intensively than the previous two sectors up to 1970, but in contrast to them they were rising consistently. After the 1971–4 boom, UoW went into a phase of decline, while SW started to grow again and, though the increase was moderate, eventually overtook GW in the number of dwellings completed.

In the Netherlands there are three forms of promotion: the private sector (*partikulieren*) (a.1 and a.2), the local authorities (d) and the non-profit-making companies (*woningbouwverenigingen*) (b). The first is made up, as in West Germany, of several different types of promotion, including private individuals or firms building houses for selling or letting on the open market, and insurance and finance companies or institutions which build for letting; the second and third kinds of promoter produce houses for letting and are analogous to the local authorities in Great Britain (see Figure 3.4).

The three sectors follow similar trends: a single cycle, of varying size which, in all three sectors, features numerous short cycles in the ascending and descending phases. The three cycles are staggered, in the sense that the high-point is reached at different times in each case. The private sector, the largest in size, reached its high-point in 1974; the non-profit-making companies in 1973; the local authorities in 1967. In contrast to the other two kinds of promoter, the non-profit-making companies showed signs of revival after 1979 and by 1981 they were

Figure 3.4: Netherlands Dwellings Completed by Form of Provision 1946-82

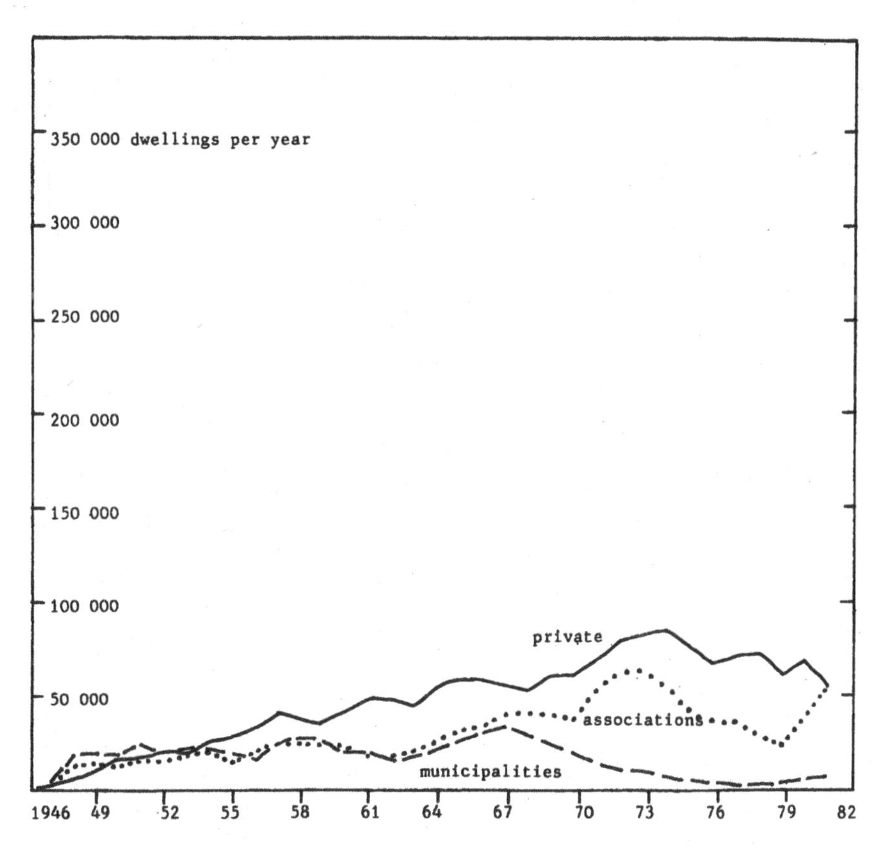

producing more than the private sector (this was the only case observed in Europe where there was a strong recovery on the part of social or public producers of rented accommodation, although by 1983 they were again in decline). The local authority cycle seems, on the other hand, to be complete.

In Sweden there are three principal forms of promotion: private (a.2), semi-public bodies (c) and co-operatives (b). Trends in the three sectors are similar to those observed in the Netherlands: a single cycle in each of the three sectors, peaking in different years. The private sector reached its high-point in 1975; the semi-public bodies in 1971; the co-operatives in 1965 (see Figure 3.5). In all three cases the upward phase of the cycle is very long, since, in contrast to what happened in the other countries, housebuilding activity in Sweden stayed high throughout the war years. In the downward phase, the proportion of

Figure 3.5: Sweden Dwellings Completed by Form of Provision 1946-82

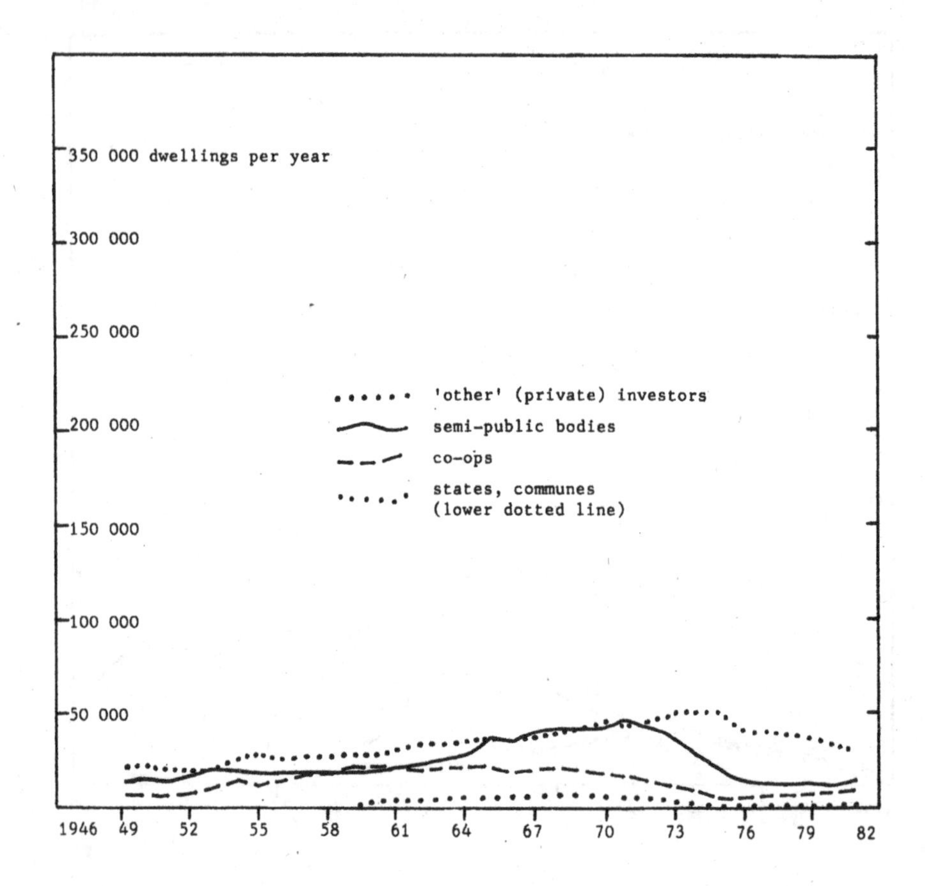

completed dwellings attributable to the various sectors changes. The public bodies sector which, in some years had been responsible for the largest contribution, fell to markedly lower percentages. The co-operative sector registered a similar fall in importance.

Data for France do not enable the same long-term comparisons as undertaken for the other four countries. A breakdown of the different kinds of private developer has existed only since 1960-2. Before then, statistics report only the total dwellings built by the private sector (with some information on different types of state financial aid to the private sector) plus the output of HLM institutions (semi-public bodies building social housing).

Topalov (1974; 1982) has suggested that there are at least two distinct types of private developer: (a.1) individuals building their own

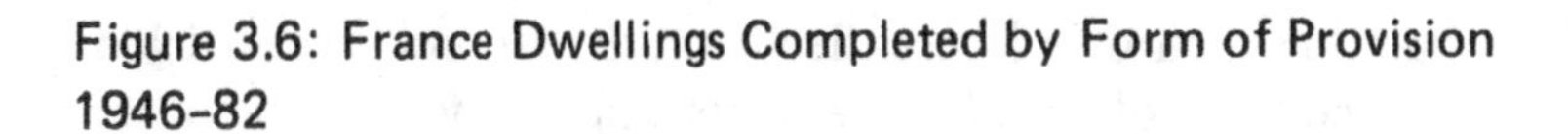

Figure 3.6: France Dwellings Completed by Form of Provision 1946–82

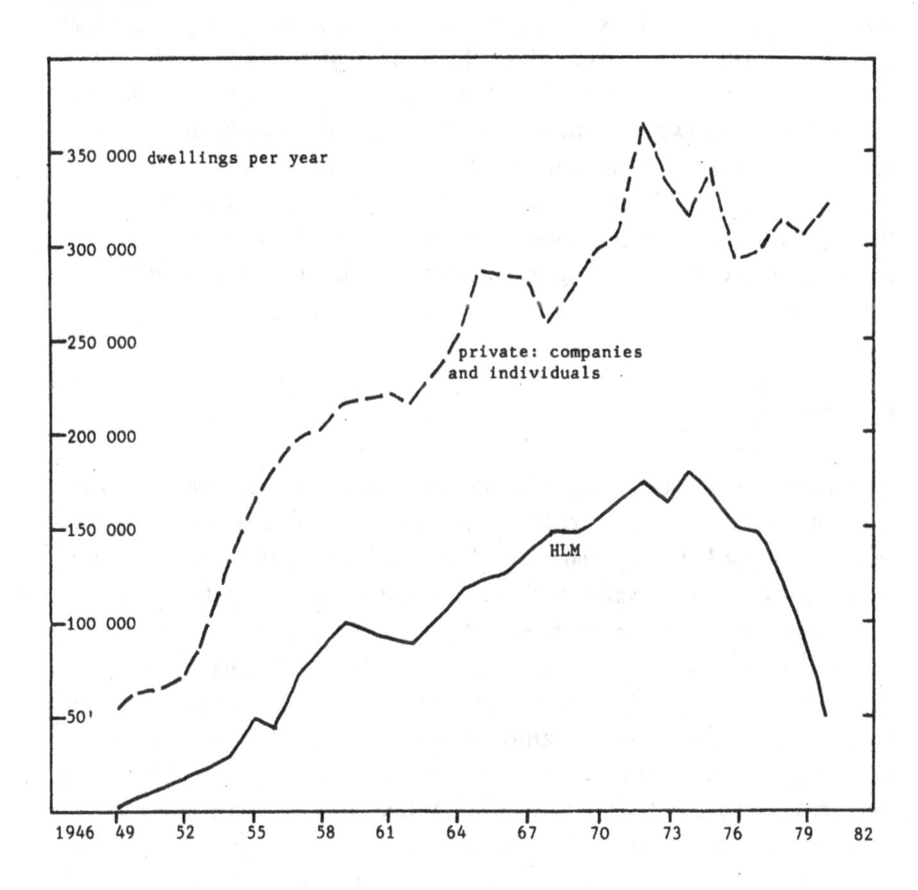

homes (usually by hiring a builder rather than undertaking the construction work themselves); and (a.2) private capitalist developers.

In terms of dwellings completed, both the private and HLM sectors grew until 1972 and 1974 respectively (see Figure 3.6). After those dates the private sector declined, with three short and irregular cycles in the downward trend, while the HLM sector collapsed. In the period of expansion up to the early 1970s private sector output grew faster than that of the HLM, and for both there were short cycles: the private sector had cyclical peaks in 1961 and 1965, and HLM in 1955, 1959, 1968 and 1972.

The relative importance of each sector has varied (Topalov 1982). Four phases can be identified: the first is from 1960 to 1964, with private developers growing in importance while both HLM and individuals building their own houses declined; the second is from 1965 to 1968, with a collapse of private developers and a growth of the other two sectors; the third lasts from 1969 to 1975, with a collapse of HLM and expansion of the other two; the fourth phase is from 1976 onwards, with private developers declining, the HLM sector collapsing and sustained growth of individuals building their own houses. What is important to note is the radical changes in the internal structure of private developers during these four phases. In the final fourth phase, in particular, there is a growing concentration of ownership of these enterprises and an increasing integration of them with financial institutions (see Chapter 2).

Conclusions

The trends described above illustrate the movement of specific forms of value increase. The short-term cyclical variations and the long-term tendencies mark the appearance of the lack of conditions favourable to the production of value and to its rotation; these conditions differ, sometimes considerably, from country to country, and require separate study. The rise or fall in housebuilding activity on the part of each type of promoter is accompanied by reorganisation processes, restructuring and internal crises. Certain kinds of promotion, or systems of social relationships as defined above, die and others are born. There are beginnings and ends. The studies of public and private promotion of housebuilding (Topalov 1982), of finance and mortgage capital and its involvement in housebuilding promotion (Frank 1982; Massey and Catalano 1978) and of the construction sector (Ball 1983; Janssen 1982) have all shown that, in the cases under consideration, periods of recovery of certain types of promotional activity may be accompanied by radical changes in the ways in which the promotion itself works and in the social relationships between the parties in the operation.

There is abundant evidence to show that, allowing for different national connotations, the forms of public and private promotion so far adopted are close, and in some cases very close, to a crisis point as far as their survival is concerned. Entire markets have disappeared and others have changed radically. In such a situation the housing needs caused by the restructuring processes and crises in national economies would seem to call, not so much for the extension of the present forms, but rather for their transformation into new forms of house production.

Notes

1. This paper is the first, initial result of a research project being carried out at the DAEST, University of Venice. The project is a comparative research on housing policies and forms of housing provision in Europe since 1946. The research is being undertaken in conjunction with Bernardo Secchi (Politecnico, Milan) and Stefania Potenza (DAEST, Venice). I should like to thank my fellow research collaborators, A. Marson, M. Savoia, C. Mazzoneni and C. Uberti, for their help in collecting data and in the formulation of the graphs.

2. On the different concept of class in Marx and in Weber, see Saunders (1980: 66–102).

3. A mechanical relationship between housing shortage and quantitative production is generally supposed with reference to the 1950s and 1960s (Heidenheimer, Heclo and Teich Adams 1975; UN 1980).

4. The UN report makes an attempt at estimating housing needs in 17 European countries. Total housing shortage estimated at the end of 1947 with reference to all 17 countries is of 14,245,000 dwellings (representing 21.9% of the pre-war building stock), of which only 3,111,000 were due to war destruction; the greatest need was for relieving overcrowding (4,017,000), for replacing unhealthy dwellings (6,517,000), and for replacing unsafe dwellings (600,000). So the total need for dwelling improvement in the 17 countries was 11,134,000, that is 79% of the total shortage. Obviously these figures differ in each country (UN 1949: 12–14). On different systems for estimating housing needs and on their subjective and ideological nature, see Godard and Pendariés (1979).

5. What Myrdal says about the importance of housing investments for national economies reflects exactly the choices of both housing and economic policy taken in Sweden since 1942. It is not by chance that among the members of the housing sub-committee inside the ECE there is A. Johansson, one of the principal authors, with Myrdal, of Swedish policy (Headey 1978: 66–78).

6. The political meaning of the main changes occurring in housing policies in many European countries and in the US, during the 1970s, has been examined in a number of studies (Ball 1982; Harloe 1981; Marcuse 1982; PEHW 1975, 1976) and discussed in international conferences such as: Istituto Gramsci-Sezione Veneta, *La casa e la sinistra in Europa*, Venezia (1980) (Folin 1982b); ISA, *Economic Crisis and Urban Austerity*, New York 1980; Hochschule fur bildende Kunste, *Krise der Wohnungspolitik*, Hamburg (1981) (Frank and Harms 1981).

7. So Barbon continues: 'Houses of the same conveniency and goodness are of more value in *Bristol, Exeter* and *Northampton*, then in the little Villages adjoining' (Barbon 1685: 20). With identical words R. Hurd has described the same phenomenon, more than 200 years later, with reference to the US: 'The best residence land in cities of 100,000 to 200,000 population runs from $75 to $150 per front foot, in cities of 200,000 population to 400,000 population from $300 to $500 per front foot, and in New York from $2,000 to $5,000 per front foot on the side streets, and $6,000 to $9,000 per front foot on Fifth Avenue' (Hurd 1903: 142). See also for the German situation in the same period R. Eberstadt (1910). The arguments by Hurd have been sustained again in the late 1950s by Wendt (1957), in opposition to Ratcliff (1949), with reference to the relationship between urban growth and land values.

8. This is the definition given in the Town and County Planning Act 1971 (Section 22(1)).

9. At the beginning of modern town planning, town development and development processes as land development are considered together (RIBA 1911: 247–332, 663–9).

10. It is the merit of Marxist studies on rent to have clarified these relationships. The most recent and complete theoretical systematisation is by Topalov (1981); see Bibliography.

11. On the behaviour and structure of different kinds of private developers see Topalov (1974, 1982); with reference to the English situation see Harloe, Issacharoff and Minns (1974), Drewett (1973), Pahl and Craven (1975), and Ball (1983).

12. 'The period since the Second World War has been characterized, unlike preceding periods, by an uninterrupted rise of land prices' (UN 1973: 74).

13. 'Land in an area planned for a satellite town for Munich was DM 0.93/m²; but when the news of the plan reached the public, the price rose to DM 250/m² – an increase of 270 times' (UN 1973: 82).

14. 'The rapid increase of land prices within the radius of the central town shows that with expansion of the urbanization process, differences between prices in peripheral areas and exterior districts of a large town become smaller in comparison with those in the central districts. Again, the rate of increase is higher in low-priced areas than in already expensively priced districts' (UN 1973: 88).

15. 'In almost all countries the share of land in housing costs has increased. Only in the Netherlands is there no change in the role of land in housing costs – even in comparison with the pre-war period' (UN 1973: 89).

References

Balchin, P.N. and J.L. Kieve (1977) *Urban Land Economics*, MacMillan, London

Ball, M. (1981) 'The development of capitalism in housing provision', *International Journal of Urban and Regional Research, 2,* 145–75

—— (1982) 'Housing provision and the economic crisis', *Capital and Class, 17,* 60–77

—— (1983) *Housing Policy and Economic Power: the Political Economy of Owner Occupation*, Methuen, London

Barbon, N. (1685) *An Apology for the Builder: or a Discourse Showing the Causes and the Effects of the Increase of Building*, Cave Pullen, London

Bowley, M. (1966) *The British Building Industry: Four Studies in Response and Resistance to Change*, Cambridge University Press, Cambridge

Donnison, D. and C. Ungerson (1982) *Housing Policy*, Penguin Books, Harmondsworth

Drewett, R. (1973) 'The Developers: Decision Process' in P. Hall *et al.*, *The Containment of Urban England*, Allen and Unwin, London, 2, 163–93

Eberstadt, R. (1910) *Das Wohnungswesen und die Wohnungsfrage*, Gustav Fischer, Jena

Folin, M. (1981) 'Crisis of public housing in Europe' in H. Frank and H. Harms (eds.), *Krise der Wohnungspolitik*, materials prepared for the European Workshop held at the Hochschule für bildende Kunste, Hamburg, 36–46

—— (1982a) 'Bisogni abitativi e lotte per la casa in Europa', *Laboratorio Politico, 1,* 175–83

—— (1982b) *Esiti della politica socialdemocratica della casa in Europa*, Angeli, Milano

Frank, H. (1981) 'Influenza del credito immobiliare sulla politica della casa' in M. Folin (ed.), *Esiti della politica socialdemocratica della casa in Europa*, Angeli, Milano

—— and H. Harms (1981) *Krise der Wohungspolitik*, materials prepared for the European Workshop held at the Hochschule für bildende Kunste, Hamburg

Godard, F. and J.R. Pendariés (1979) *Les modes de vie dans les discours de la représentation: Institutions locales et production politique des besoins*, Laboratoire de Sociologie de l'Université de Nice, Nice

Harloe, M. (1981) 'The recommodification of housing' in M. Harloe and E. Lebas (eds.), *City, Class and Capital*, Edward Arnold, London, 17–50

——, R. Issacharoff and R. Minns (1974) *The Organization of Housing: Public and Private Enterprise in London*, Heinemann, London

—— and M. Martens (1984) 'Comparative Housing Research', paper presented at SSRC Conference *Housing Research in Britain: the Next Decade*, Bristol, 1983. *Journal of Social Policy*, forthcoming

Harvey, J. (1981) *The Economics of Real Property*, MacMillan, London

Headey, B. (1978) *Housing Policy in the Developed Economy*, Croom Helm, London

Heidenheimer, A.J., H. Heclo and C. Teich Adams (1975) *Comparative Public Policy: the Politics of Social Choice in Europe and America*, MacMillan, London

Hurd, R. (1903) *Principles of City Land Values*, The Record and Guide, New York

Janssen, I. (1982) 'L'organizazione del capitale nella produzione edilizia e lo sviluppo dell'edilizia residenziale', in M. Folin, *Esiti della politica socialdemocratica della casa in Europa*, Angeli, Milano, 155–64

Kalecki, M. (1944) 'Three ways to full employment' in *The Economics of Full Employment*, Basil Blackwell, Oxford, 39–58

Lichfield, N. and H. Darin-Drabkin (1980) *Land Policy in Planning*, Allen and Unwin, London

Marcuse, P. (1982) 'Determinants of State Housing Policies: West Germany and the United States' in S. and N. Fainstein (eds.), *Urban Policy under Capitalism*, Sage, Beverly Hills, 83–115

Martin, A. (1979) 'The Dynamics of Change in a Keynesian Political Economy: The Swedish Case and Its Implications', in C. Crouch (ed.), *State and Economy in Contemporary Capitalism*, Croom Helm, London

Massey, D. and A. Catalano (1978) *Capital and Land: Landownership by Capital in Great Britain*, Arnold, London

Merrett, S. (1979) *State Housing in Britain*, Routledge and Kegan Paul, London

Nycolaas, J. (1974) *Volkshuisvesting*, Socialistiese Uitgeverij, Nijmegen

Pahl, R.E. and E. Craven (1975) 'Residential expansion: the role of the private developer in the South-East' in R.E. Pahl (ed.), *Whose City?*, Penguin Books, Harmondsworth, 105–23

PEHW (1975) *Political Economy and the Housing Question*, Political Economy of Housing Workshop, Conference of Socialist Economists, London

—— (1976) *Housing and Class in Britain*, Political Economy of Housing Workshop, Conference of Socialist Economists, London

Ratcliff, R.U. (1949) *Urban Land Economics*, Greenwood, Westport

RIBA (1911) *Town Planning Conference: London, 10-15 October 1910: Transactions*, RIBA, London

Saunders, P. (1980) *Urban Politics*, Penguin Books, Harmondsworth

Topalov, C. (1974) *Les promoteurs immobiliers: Contribution à l'analyse de la production capitaliste du logement en France*, Mouton, Paris

—— (1981) *Le Profit, la Rente, et la Ville: Eléments de Theorie*, Centre de Sociologie Urbaine, Paris

—— (1982) 'Transformazione dei sistemi di produzione della casa e politiche statali (1950–1978)' in M. Folin (ed.), *Esiti della politica socialdemocratica della casa in Europa*, Angeli, Milano, 73–120

UN – ECE (1949) *The European Housing Problem*, UN, Geneva

—— (1954) *European Housing Progress and Policies in 1953*, UN, Geneva
—— (1955) *European Housing Developments and Policies in 1954*, UN, Geneva
—— (1956) *European Housing Progress and Policies in 1955*, UN, Geneva
—— (1957) *European Housing Trends and Policies in 1956*, UN, Geneva
—— (1958) *European Housing Trends and Policies in 1957*, UN, Geneva
—— (1959) *European Housing Trends and Policies in 1958*, UN, Geneva
—— (1960) *European Housing Trends and Policies in 1959*, UN, Geneva
—— (1961) *European Housing Trends and Policies in 1960*, UN, Geneva
—— (1963) *European Housing Trends and Policies in 1961 and 1962*, UN, New York
—— (1973) *Urban Land Policies and Land-Use Control Measures: Volume III Western Europe*, UN, New York
—— (1980) *Study on Major Trends in Housing Policy in EEC Countries*, UN, New York
Wendt, P.E. (1957) 'Theory of Urban Land Values', *Land Economics, 33*, 228–40
—— (1963) *Housing Policy – The Search for Solutions: A Comparison of the United Kingdom, Sweden, West Germany, and the United States since World War II*
Wicksell, K. (edn 1954) *Value, Capital and Rent*, Allen and Unwin, London

4 LAND RENT AND THE CONSTRUCTION INDUSTRY

Michael Ball

Introduction

Does land rent affect the production process of the construction industry? This question produces clear theoretical divisions over urban land rent. In this chapter I should like to present my own view on the matter and place it in the context of the wider debate. As ever, the context must come first.

Theories of Land Rent

On the issue of the influence of land rent on the production process in the construction industry neoclassical economics is at best agnostic: it has never been proved to exist; yet, on the other hand, it has never been proved not to exist either. The whole issue is irrelevant to the objectives of the theory so, not surprisingly, little thought has been given to it. For neoclassical economics the interesting question about rent is its role as an allocative mechanism. Rent and the land market help to determine the spatial ordering of land uses; use rather than production and use is the order of the day. Building techniques are pre-given in neoclassical models of urban land rent; so the parameters of the productions function are exogenous, if they are ever examined at all (Muth 1969; Richardson 1977; Needham 1981).

In the usual neoclassical approach to the topic of rent, land rent can at most have an indirect effect on the construction industry, by altering the intensity with which land is used and thus the type of buildings built. A high-rise office block obviously requires different production methods from low-rise housing, and they both present different potentials for innovation. But it is the physical nature of the building not the economic appropriation of rent that really creates the effect on technology.

Neo-Ricardian approaches to land rent produce similar conclusions on the rent question to those of neoclassical economics (Scott 1980).

This is not surprising given the common origins of the two schools of rent theory in Ricardian economics. Again the array of potential building techniques is known prior to the determination of rent.

Only Marxist theory has raised seriously the possibility that land rent may influence production processes in the construction industry (cf. Ascher 1974 and Lipietz 1974). Unlike other theories of rent, the Marxist approach argues that contradictory social relations are at the centre of the analysis of economic mechanisms within capitalism. Production processes are the outcome of class struggle in production between capital and the working class over the extraction of surplus value (Braverman 1974; Wood 1982). Land rent is an intervention by landed property into that struggle over production: one which may change its outcome. The intervention of landed property produces significant change when it alters the profitability or feasibility of specific production methods or the location of production. By doing so, it also intervenes in the competition between capitals (and possibly with non-capitalist producers) as well as between classes. That competition could be over the location of production, which in the case of the building industry will simultaneously be over the location of its products. The influence of rent on the spatial distribution of economic activities consequently cannot be separated from its potential influence on the immediate process of production. In the case of urban areas this means that the effect of land rent on the construction industry cannot be ignored. Where the activities using the structures produced by the building industry are themselves operating on capitalist criteria there is a double influence of rent: on the production process going on within buildings as well as on the production and cost of built structures themselves.

How can rent affect capitalist production? The impact on production results from the monopoly ownership of a plot of land that is inherent in the notion of landed property and the need to pay rent (or its capitalised form as a land price) to overcome the barrier to use created by landownership. The effect on production of the surplus value appropriated as rent is not simple and the outcome, when significant, can take various forms: (i) through reducing the surplus value available for investment (Massey and Catalano 1978); (ii) by lowering the profitability of investment (Ball 1977; Fine 1979); or, finally, (iii) by altering the terms and conditions under which capital investment can be made (cf. Chapter 6 by Ben Fine). The object of the analysis of rent, therefore, is to investigate the significance of rent on production; only in this way can the allocative and distributional consequences of rent be clearly understood. For urban development this means that the impact of rent on the construction industry cannot be ignored.

Before proceeding to examine construction, a few additional comments should be made about Marx's theory of rent in general. It is important to be aware of the contingent nature of the intervention by landed property. It depends on the land use in question and on the moment in history. Moreover, the intervention may be beneficial for capital accumulation or it may not. Similarly the intervention might have no effect on the content of production at all. An instance of this in Marxist rent theory occurs with the notion of differential rent I, where it is assumed that equal amounts of capital are invested on each land and that the way in which production is undertaken depends simply on differences in 'fertility' and location (although the likelihood in practice of the investment of equal amounts of capital on different lands is small). At other times, as is often claimed for the construction industry, the impact of rent on methods of production can be substantial.

Each possible intervention of land rent in production has to be examined through theoretical and empirical analysis to find out its effects. As Marx said early on in his examination of rent in *Capital*:

> the price of things which have in themselves no value, i.e. are not the product of labour, such as land, or which at least cannot be reproduced by labour, such as antiques and works of art by certain masters, etc., may be determined by many fortuitous combinations. In order to sell a thing, nothing more is required than its capacity to be monopolised and alienated. (*Capital III*, 633)

'Fortuitous combinations' do not, of course, mean random combinations. Theoretical analysis can highlight the conditions that structure the situations in which such 'fortuitous combinations' arise. Marx's analysis of ground rent in *Capital* follows this procedure, examining the impact of rent via the categories he classified as differential rent, types I and II, and absolute rent.

The theory of agricultural rent elaborated in *Capital* can be applied to the extraction of other commodities from the land, like timber and minerals. Commodity production in factories is influenced in a slightly different way as Marx explains in his introductory chapter on differential rent (see Ball 1977). Although Marx carefully brings spatial differentiation into his analysis, the effects on industrial production of the high rents arising from competition for land in urban areas is not considered. As regards other urban activities even less is said, and there is nothing on construction at all.

The omissions by Marx in his analysis of rent have a simple explanation but their existence has caused considerable subsequent confusion, so it is worth briefly pointing out the reasons for the partial nature of Marx's examination of rent. As with so much of his writing on rent, when looking at factory production (and agriculture), Marx was primarily interested in the effect of rent on the production of the same commodity at different locations rather than with competition for the same land by producers of different commodities. He regarded competition between capitals in different industries as being of second order to the determinants of accumulation in one industry. The key question about rent, therefore, was its affect on accumulation in an industry. Inter-industry competition would then be structured by the outcome of accumulation in each industry itself. Only when considering the effect of absolute rent did Marx examine in any depth the effect of rent on competition between industries (via its impact on the equalisation of the rate of profit) (see Fine 1979). Unfortunately, Marx did not live to write the work on competition he planned; with it was lost any comprehensive discussion of rent and the competition for land between different potential uses. Urban rent consequently remained an empty box.

Recent Marxist writers have tried to fill the 'urban' gap by applying the concepts of Marx's theory of agricultural rent directly to the urban situation (Lipietz 1974; Ascher 1974; Edel 1977; Harvey 1973). The transposition has not been particularly successful, except as a heuristic device directing attention away from the smooth optimising inherent in the neoclassical economics approach towards recognition of the multitude of conflicts and contradictions that arise over urban space.

The major problem faced by attempts to transfer the concepts of agricultural rent to the urban context is that the economic mechanisms through which rent is appropriated are not the same. The basic points associated with the monopoly conferred by landownership still hold but the conditions structuring the 'fortuitous circumstances' are different.

The mechanisms elaborated for agriculture in Marx's famous categories of rent are based on the production of a single uniform commodity at different locations. A theory of the determination of market value (by the price of production on the marginal land) and an analysis of differences between prices of production on individual land plots and the market value determining price of production enable a theory of differential rent to be elaborated. Subsequent examination of rates of investment on each land, the differential rent II cases, completes the analysis of differential rent (Ball 1977) from which Marx proceeded to absolute rent, considering the effect of monopoly along the way.

The notion of a uniform commodity and a single market price cannot hold in the urban context. Ground rent is charged on buildings which by definition are all different, if for no other reason than their location. Selling prices for built structures similarly vary considerably. The effect of competition between building producers and between them and owners of existing structures tends to sharpen those differences in price by highlighting the uniqueness of each building (because of its location if nothing else), rather than to obliterate such differences in the selling prices of the same commodity as tends to happen in agricultural markets. Of course, there are parallels with the agricultural case in the determination of urban rent. The advantages of centrality for many activities and the existence of transportation costs create the possibility of differential rents arising. Whereas the cost of creating new built structures (and their location) will influence the rents that can be charged on existing buildings (echoes of the marginal land). But none of these points is sufficient to explain the pattern of rents in any urban area. The 'fortuitous circumstances' are constrained in different ways from those in agriculture.

One other tack has also been adopted in trying to use Marx's categories of agricultural rent for urban situations. Attempts have been made to apply the concept of absolute rent in explanations of the 'backward' nature of the forces of production in the building industry. It has been argued that absolute rent exists because a low organic composition of capital in the building industry enables landowners to extract this type of rent from the surplus value produced in the building process. Without this intervention of landed property, the surplus value appropriated as absolute rent would be redistributed to higher-organic-composition sectors in order to equalise the rate of profit between industries. The appropriation of absolute rent, furthermore, is said to act as a barrier to investment in the building industry, thereby restricting the development of more productive techniques with higher organic compositions of capital (Ascher 1974; Lipietz 1974). Absolute rent therefore is both an effect and a cause of the low organic composition of capital in the building industry.

Whether or not construction does empirically have a low organic composition of capital (which is contentious once circulating constant capital is included), the argument about absolute rent and the building industry, as it stands, is circular. Marx avoided circularity in his analysis of absolute rent in agriculture by making one consequence of the monopoly ownership of land a central and necessary part of his argument. The existence of landed property, he deduced, would result in a demand for rent on land which would otherwise pay no (differential) rent. He theorised, in other words, about the historically specific conditions under which the monopoly power of landed property could generate absolute rent (i.e. the existence of uncultivated land which would feasibly be brought into cultivation on which no rent was currently paid). In this way, he avoided the circularity of simply relying on organic composition alone.

Direct transposition of Marx's argument on absolute rent in agriculture to the urban context is obviously impossible. Whilst it might be feasible, particularly in the nineteenth century, to posit the potential existence of a margin of cultivation beyond which land paid no rent, in modern urban contexts competition between land uses ensures that rent will never be zero. The logical contradiction between the concept of private landownership and zero rent stressed by Marx in his discussion of absolute rent in *Capital* and in his critique of Ricardo in *Theories of Surplus Value* seems, therefore, to have little relevance for the urban situation.

Land Rent and the Construction Industry

When looking at the effect of land rent on the construction process, it is important to start with the social relations associated with building rather than with categories of rent that are predicated on a prior knowledge of the social relations in question. The type of construction industry in existence and the social organisation of its subsectors depends on the contemporary mode of production and the historical development of class relations in the industry (Ball forthcoming; Chapter 4). The theoretical ordering of well-known, if historically specific, social agents associated with building takes us a long way towards understanding the effect of land rent on the industry.

The construction industry occupies a key place in the creation of the built environment of urban areas. To state the obvious: buildings get built in the building industry. But why and how they get built cannot be explained by looking at construction alone. The building process is part of a specific structure of building provision, one that helps to determine the social relations of production as well. A structure of building provision constitutes the social and physical framework of production, exchange and consumption associated with the social provision of a building type (Ball forthcoming; Ch. 1). The social organisation of building provision is historically specific, so structures of provision have to be specified for particular types of building and their contemporary forms of production, consumption and exchange.

Examples of structures of building provision in modern Britain are owner-occupied housing provision and council housing provision using private building contractors (Ball 1983 a and b). As Marino Folin has noted in the previous chapter, other countries have different structures of provision for the same types of housing although there might be strong similarities (see, in particular, the French example of owner-occupation discussed in Chapter 2 by Christian Topalov, especially how the position of social agents and, therefore, the nature of owner-occupied housing provision changed over time).

Landownership is one of the social relations involved in a structure of building provision. Its influence on production, therefore, depends on the dynamic of the overall structure of provision. Consequently, there is unlikely to be one singular effect of land on construction; instead the effect depends on the contemporary situation facing individual types of building provision.

A hierarchy of analysis is required to evaluate the effect of rent on the construction industry. First, the social relations associated with particular types of building provision need to be specified. Then the effect of landownership can be evaluated as one of the constituents of a structure of provision. The effect of landownership, moreover, is likely to vary over time as the particular structure of provision changes through external pressures or under the dynamic of its own internal contradictions.

Before looking at some specific types of building provision, it is worth initially pointing out a consequence for Marxist rent theory of the argument that it is impossible to have a singular influence of rent on the construction process. An attempt to transpose the rent mechanisms elaborated by Marx from agriculture to construction would have to deny the complexity and variety of the social relations associated with building provision. Such a use of one part of *Capital*, paradoxically, has therefore to deny the theoretical foundations of the rest: that is, the primacy of historically determinate class relations.

Some empirical examples can illustrate and enable further elaboration of the general points being made. The two basic forms of capitalist building, building to contract and speculative building, will be considered. Some theoretical deductions about the likely impact of rent can be made once the general nature of these two forms of building is specified, but only examination of particular historical circumstances can bring out the actual effect of the rent relation.

Modern Contract Building

Contracting refers to the situation where a capitalist construction firm builds a pre-specified structure for a known client. The 'client' can be one of many social agents. It may, for example, be the state, a property developer, a capitalist enterprise that wants to use the structure as an owner occupier or a private individual. What is important for the discussion here is the relationship between building capitalist and landowner. And the link is obviously a tenuous one. The contractor has no legal or financial relationship with the land or its owner prior to the project. The building firm simply tenders for the work and undertakes it when the tender is successful, once the 'client' has gained control (through renting or purchase) of the land site. The client bears the burden of land costs. There might be indirect effects on the building construction process via the effect of land costs on the location of projects, the clients' willingness to instigate building projects and the frequency with which they do so. There, nevertheless, is no possibility of rent directly affecting the profitability of building production, or accumulation by the building firm, or the building techniques used. The potential indirect effects of rent, moreover, are only one of a whole ensemble of influences. They are likely at most to be subsidiary to the general influences of the level of economic activity and the state of credit availability. The external environment as a whole has far less effect on the nature of construction processes under contracting than the class relations associated with production and exchange that have developed historically with building to contract (Ball forthcoming: Ch. 4).

The essential point about this summary of contract building is that no clear economic mechanism can be specified by which the appropriation of rent necessarily has an effect on the construction process. In fact, the reverse is likely to hold: the situations where land rent affects construction are likely to be highly exceptional. Detailed historical research and argument are necessary to discover such exceptional situations, theory alone cannot help.

Speculative Building

Speculative builders build for a general market rather than for known clients; as a result they have to make calculations about what will sell and about the timing, location and scale of their developments. All that this definition of speculative building has done so far is to specify the market context (i.e. exchange relations) in which speculative builders operate. A number of combinations of social agents are possible with speculative building depending on the actual development of class relations within this form of building. The land development and building process may be carried out by separate capitalist agencies, for example, as they generally were in nineteenth-century Britain, or they may be combined under the control of one capital as occurs with present-day British owner-occupied housebuilding.

Generally, speculative building necessitates the acquisition of land by an agency in the building process well before the buildings are built or sold. When building firms purchase or lease land there is likely to be some influence of land on the production process. Builders obviously acquire land because it is profitable to do so; they undertake production for the same reason. The two determinants of profit may harmoniously combine or come into conflict. *A priori* the chance of conflict seems high. Economic mechanisms associated with land and production can be explored to see whether that hunch is correct. As those economic mechanisms depend on the particular nature of the relations between social agents involved in the building process, once again they are predicated on the social relations associated with structures of building provision. Examples from the development of speculative housebuilding in Britain illustrate some of the potential influences of land rent on construction. Distinct structures of provision are involved, so the effect is not uniform.

During the second half of the eighteenth century and in the early years of the nineteenth century, capitalist relations of production came to dominate the British building industry (Cooney 1955; Clarke 1982). Speculative building of private housing to rent emerged as an important component of the new capitalist building industry (Ball 1981). Contemporary circumstances led to a social division within housing provision between landowners, land developers and capitalist builders. Generally landowners sold or leased land to developers, who laid out and serviced the building estates, the subdivided plots of which were purchased or leased by speculative builders, who in turn leased or sold the houses they built to housing landlords.

What were the effects of rent relations on housebuilding in this structure of provision? In the first place, the ability to appropriate land development profit and rent from housing provision encouraged the emergence of this form of provision. The lure of land profits helped to break down the petty commodity production associated with pre-capitalist craft builders. Whereas the periodic economic crises which befell this structure of provision helped to create a pliant, impoverished building workforce, shorn of its earlier guild rights. The attractions of the enormous profits that could be made from the early land development stages of housebuilding encouraged the large scale mobilisation of capital necessary for housebuilding. In the absence of any significant state infrastructure expenditure, capitalist urban development would have been literally impossible without that mobilisation.

The ability to appropriate substantial rents from urban development consequently had beneficial influences on the capitalist form of housebuilding that emerged in early nineteenth-century Britain. There might be conflicts over the shares of land development profits going to either the builder, the land developer or the initial landowner but the existence of the mass of profit from which the individual shares were derived was determined by the capital induced to invest in housing development. Put another way, mass production and capitalist-style divisions and control of labour came to housebuilding with the help rather than the hindrance of landed property. (Clarke and Janssen (1983) and Cannadine (1981) provide some useful detailed descriptions of the role of the landed aristocracy in nineteenth-century urban development.)

Besides these positive effects of the existence of landed property on accumulation in the building industry, there were also negative ones. They were associated with all three types of barrier to accumulation mentioned earlier (i.e. reduced mass of surplus value, lower profitability of investment and inhibiting terms and conditions of investment). A consequence, for example, of the power of the independent development agencies associated with urban building was that the actual productive enterprises involved in speculative building remained small and were comparatively weak in this structure of provision. They were in no position to revolutionise the forces of production in housebuilding in ways similar to those being achieved by capital in other industries, including, it should be noted, ones where land rent was a significant outlay for the capitalist, as in agriculture and mining (see Chapter 6). Land development and the enhanced ground rents it created attracted most of the capital and entrepreneurial talent in housebuilding. Particularly successful builders would tend to drop their building operations and become either housing landlords or estate developers themselves.

Much of the enhanced ground rent swallowed up the potential profits of building development, including the surplus value arising from the capitalist social division of labour in construction. Potential profits from innovation in building techniques, therefore, would quickly be appropriated through ground rents once they became commonly used. Additionally, the terms and conditions of building leases inhibited considerably the ability of builders to alter the timing and content of their operations (Ball 1981).

The most famous case of a building capitalist turning estate developer is Thomas Cubitt (Hobhouse 1971). His operations in London after 1830 also illustrate another consequence of the social relations of that structure of building provision. Because of the barriers to investment faced by builders in early nineteenth-century speculative housebuilding, the quality of building was frequently poor. In situations where high quality construction was required, estate developers had either to co-opt building contractors to their development ventures or build themselves, using directly employed workers, rather than sublet plots to individual speculative builders. Cubitt adhered to these principles: using his own workforce to build Belgravia for the aristrocracy, whilst subletting plots for middle-class Pimlico. The difference in construction quality still shows today.

Later in the nineteenth century, the negative effects of landownership on speculative housebuilding began to outweigh the positive ones. Capitalism was by then firmly established. Private landownership and the gains from urban rent were no longer required to force subordination to capital on a reluctant workforce. The state, both locally and nationally, began to invest substantially in the infrastructure necessary for urban development, as did a series of capitalist agencies associated with public transportation and service utilities. Instead of being a means by which capital was channelled into urban development the landowner-estate developer partnership appropriated the development gains, created by the actions of the state and the utilities, that would otherwise have gone to the builders. (Michael MacMahon deals with other impediments to development caused by the contemporary property rights of private landownership in Chapter 5.)

The difficulties facing building capital in trying to raise labour productivity substantially through increases in the scale of their operations and through investing in new methods of production eventually brought the structure of provision into profound economic crisis in the early twentieth century (Ball 1983a). It also gave Britain an enormous legacy of slums and dilapidated, sprawling suburbia.

The parallels with the effect of landed property on nineteenth-century British coalmining, as described by Ben Fine in Chapter 6, are considerable. In both industries capital gained from the existence of a powerful landowning interest during the early stages of capitalist development. Yet increasingly throughout the nineteenth century productive capital and landownership came into conflict over accumulation through innovations, or their absence, in production. The parallel ends after the First World War, however: unlike the coal industry, speculative housebuilding was rejuvenated without requiring the destruction of the landowning interest through nationalisation.

The historical conditions in which owner occupation grew into a major tenure in the nineteen twenties and thirties were sufficient for the revitalisation, at least temporarily, of speculative housebuilding. Productive building capital was able to compete the increasingly parasitic estate developer out of existence, whilst agricultural crisis put suburban landowners in a weak position. In the space of a few years giant housebuilding firms emerged; the largest building as much as 1,000 times the output levels of their pre-1914 predecessors. Enormous development profits could be made in the inter-war owner-occupied housing boom, little of which were appropriated as rent so the bulk remained to further the capital accumulation of the housebuilder (Ball 1983a; Ch. 2).

The transformation of owner occupation into the major tenure in Britain since 1950 has created a continued demand for housing produced by the speculative housebuilding industry. By 1984 over 60 per cent of all British households were owner occupiers compared with less than 30 per cent after the Second World War. A number of trends, therefore, have benefited speculative housebuilding capital. But others have not – including a revitalised strength of landownership. Rather than go through a detailed description of the post-war events, I should like to complete the analysis of speculative housebuilding by presenting a framework through which those events can be understood.

Speculative Housebuilders, Landowners and Building Workers

In Britain speculative builders generally control all of the functions involved in converting a land site from its previous to its new use. There is rarely an intermediate land dealer or developer who sells serviced plots with planning permission to builders. Instead, when acquiring land, builders deal directly with landowners and their objective is to minimise the conversion of the profit made on development into land rent as constituted by the land price paid to the landowner. In this sense speculative builders are akin to capitalist tenant farmers who also try to minimise the appropriation by landowners of potential profit as ground rent. But speculative builders differ from farmers in terms of their product and its market, so the mechanisms used to resist the appropriation of ground rent differ. The actions taken by speculative builders over the conversion of profit into rent are important in understanding the nature of this section of the construction industry because they affect the production methods used and the way the workforce is employed.

Speculative builders are a type of commercial capital buying cheap and selling dear, and they profit from getting the timing of those operations right. Much better selling prices obviously exist during house buying booms than during slumps. Timing is important for the purchase of land as well and there is no necessary reason why the best time to sell completed houses coincides with the best time to buy land (the reverse is, in fact, more likely). Speculative builders as a result hold land banks. Stocks of development land enable them to produce at the best times, to cut back production during market slumps and to acquire attractive sites when they come onto the market.

As well as being a type of commercial capital, speculative housebuilders are also a type of industrial capital involved in the productive act of housebuilding. The circuits of commercial and industrial capital, with the frequently diverging imperatives they imply, exist as a contradictory unity within the speculative builder's operations. Within the overall circuit of capital the requirements of production are subordinated to the needs of the speculative timing of purchases and sales. The subordination arises because the turnover of capital for a speculative builder does not depend on steady production rates but on the successful manipulation of land purchases, development programmes and building sales. Given the volatile market for owner-occupied housing it is highly unlikely that steady output levels can be achieved, so the production methods used have to facilitate variable output levels. The increasing volatility of property markets since the end of the long post-war boom have heightened the need for flexibility in production. As a result small runs of standardised units based on simple, repetitive building tasks carried out by a casualised workforce characterise this sector.

The impact of the development side of speculative housebuilding on the process of production has been considerable. Building techniques have to be able to accommodate generally low rates of output on each site, with periodic sharp variations in the output required. Casual employment conditions similarly reflect the importance of the variability of production. Specific forms of management control of the workforce stem from these employment conditions: piecework rates of pay, for instance, impose a self-discipline on the pace of work of individual workers. This is particularly important as they undertake physically fragmented tasks not amenable to detailed managed supervision. Once again, the physical fragmentation is more the result of the building techniques used than of the inherent nature of housebuilding. Taking these points together, it can be concluded that the rent relation between speculative housebuilder and landowner has been one of the determinants of the economic framework structuring the struggle over production between housebuilding firms and building workers.

Although housing development gain is divided into the builder's development profit and the landowner's land price, the exact division between the two components is theoretically indeterminate as it depends on the contemporary balance of power between builder and landowner (Ball 1983a; Ch. 5). Moreover, it is important not to be mesmerised by the conditions structuring the share of surplus value going to the builder and the landowner. The distributional struggle over rent through its effect on production also helps to determine the mass of surplus value out of which the shares are divided. Land rent, in other words, is not simply a distributional issue.

Conflicts in the land market are obviously influenced by wider contemporary economic and political forces. In Britain, recurring economic crises, inflation and the changing structure of capital away from manufacturing towards financial activities have all produced effects within the land market. The growth of financial landownership, as documented by Massey and Catalano (1978), is an indication of the longer term view now being taken of investment in land. A similar trend has occurred in the speculative housebuilding industry, where rapid monopolisation has taken place over the past decade so that the industry is now dominated by a few large producers, who themselves are frequently owned by types of capital looking for long-term investment returns (Ball 1983a; Ch. 3).

Yet changes in the residential land and housing markets are not simply reflex responses to wider social changes; conditions within those markets determine whether or not they are attractive places for these new forms of capital. Rising long-term land prices may explain why the structure of landownership in Britain is changing but that does not explain why residential and agricultural land prices rose in the first place to attract the new type of landowner. To understand the increase in residential land prices it is necessary to examine the dynamic of its structure of provision, which means looking at the social relations in it, including those associated with land. A secular rise in house prices has sustained rising land prices. Yet land prices should not be seen merely as Ricardian residuals: attempts by builders to minimise the conversion of potential building profits into land profits have influenced production methods as argued earlier. The production methods adopted are not conducive to substantial productivity increases and make it difficult to sustain a sufficiently large skilled workforce. Construction costs over the long-term rise considerably as a result, forcing up house prices. The existence of private landed property, in other words, has helped to create the conditions in the housing market which make land such an attractive investment.

Conclusions

From what was said in the previous section it can be seen that the influence of landed property on the construction industry is far from simple. Rent mechanisms can be specified, but they do not neatly replicate Marx's categories for agricultural rent. The effects of rent, furthermore, depend on the structure of building provision of which a particular sector of the construction industry is a part and on the moment in history.

Notions of simple rent mechanisms cannot encompass the complexities of the impact of landed property on the construction industry. Examples cited earlier showed cases where land rent had little or no effect on construction (in building to contract); where it had positive effects on the flow of capital into construction (early nineteenth-century British speculative housebuilding); and where in various ways it seems to have had a deleterious effect on production, contributing in some instances to an overall crisis in the structure of provision in question (e.g. rental housing provision in Britain in the years prior to the First World War). The examples cited in the paper, moreover, are only illustrative; they do not encompass all the possible consequences for the construction industry of the existence of private landed property.

To come to conclusions about the effect of rent it is necessary to look at detailed historical situations rather than to make gestures towards some grand general theory. This does not mean collapsing into the empiricism of the 'facts' of history. Even though the effects of rent depend on historical circumstances the conditions that structure the operations of landed property at those points in time still need to be theorised; analysing rent mechanisms and evaluating their consequences are part of that theorisation.

References

Ascher, F. (1974) 'CME et secteur de production du bâtiment et des travaux public (BTP)' in *Urbanisme Monopoliste, Urbanisme Démocratique*, Cahiers du CERM, Paris

Ball, M. (1977) 'Differential rent and the role of landed property', *International Journal of Urban and Regional Research, 1*, 380–403

—— (1981) 'The development of capitalism in housing provision', *International Journal of Urban and Regional Research, 5*, 145–77

—— (1983a) *Housing Policy and Economic Power: the Political Economy of Owner Occupation*, Methuen, London

—— (1983b) 'Housing production. Do we need a new research programme or a new type of housing research?', paper presented at the Social Science Research Council Conference on 'Housing Research – the next decade', September; available in mimeograph from ESRC, London

—— (forthcoming) *Building Blocks: Rethinking City Structure*, mimeograph

Braverman, H. (1974) *Labour and Monopoly Capital*, Monthly Review Press, New York

Clarke, L. (1982) 'The transition from a feudal to a capitalist mode of building production in Britain' in *The Production of the Built Environment*, 3, 1.52–1.70, Proceedings of the Bartlett Summer School, 1981, University College, London

Cooney, E. (1955) 'The origins of the Victorian master builder', *Economic History Review, 8*, 167–76

Edel, M. (1976) 'Marx's theory of rent: urban applications', *Kapitalstate, 4-5*, 100–24; also in *Housing and Class in Britain*, Political Economy of Housing Workshop, Conference of Socialist Economists, London

Fine, B. (1979) 'On Marx's theory of agricultural rent', *Economy and Society, 8*, 241–278

Harvey, D. (1973) *Social Justice and the City*, Edward Arnold, London

Hobhouse, H. (1971) *Thomas Cubitt: Master Builder*, Macmillan, London

Lipietz, A. (1974) *Le Tribut Foncier Urbain*, Maspero, Paris

Marx, K. (1968) *Theories of Surplus Value, II*, Lawrence and Wishart, London

—— (1974) *Capital, III*, Lawrence and Wishart, London

Massey, D. and A. Catalano (1978) *Capital and Land: Landownership by Capital in Great Britain*, Edward Arnold, London

Muth, R. (1969) *Cities and Housing*, University of Chicago Press, Chicago

Needham, B. (1981) 'A neo-classical supply-based approach to land prices', *Urban Studies, 18*, 91–104

Richardson, H. (1977) *The New Urban Economics*, Pion, London

Scott, A. (1980) *The Urban Land Nexus and the State*, Pion, London

Wood, S. (ed.) (1982) *The Degradation of Work?*, Hutchinson, London

5 THE LAW OF THE LAND: PROPERTY RIGHTS AND TOWN PLANNING IN MODERN BRITAIN

Michael McMahon

Introduction

Recent studies of law and administration reveal that the state has played an important role in the development of commodity relations (Corrigan 1980; Kay and Mott 1982; Scott 1980). Some of these works, however, suffer from various forms of teleology. Scott's urban land nexus, for example, is built on a projection of contemporary land market phenomena (as he understands them) back into the nineteenth century. And Corrigan's analysis, while provocative, none the less portrays state interventions as inexorably leading to an economy centred on a series of interlocking markets. The history of nineteenth-century landed property shows that the relationship between the state and landed property is more complex.

Marx was perhaps the first to indicate that the state is implicated in the contradictory nature of modern forms of landed property. He argued that the exclusive rights conferred by the extension of the sphere of private property to land not only excluded wage labourers but also functioned as a monopoly against capital. This point underlay his position that the tenurial relations which prevailed within the landed estates of his day inhibited the development of land as an element of production. It also provided him with the basis for speculating that the development of relations of landownership would involve further intervention by the state. In referring to demands for 'the breaking up of large landed properties', Marx (1973: 279) was probably referring to the contemporary agitation for legal reforms to facilitate development of a land market. The context of this statement also suggests that he saw this type of intervention as one of the ways out of the particular contradictions associated with rural landownership in Victorian England. These interventions did occur, and they did result in the break-up of the estate system. But, from what we know today, they did not eradicate the contradictions that Marx examined. Instead they transformed the contradictions between land and capital.

Over the past hundred years landed property has not only been developed as a private object of exchange; it has also been constituted as a complex object of public administration under the laws of modern town planning. The *raison d'être* for these laws is generally held to be the need for the rational and comprehensive development of land – a goal which the market cannot achieve by itself. Yet the elaboration of the state's capacity to engage in town planning has not achieved the intended effects. The proliferation of laws to constitute landed property as a use value has been accompanied by its development as an exchange value. Together they have made landed property an important form of fictitious capital (Harvey 1982) and thereby transformed the contradictory nature of landownership. Speculative trading, for example, has grown to the extent that landed property has become a central element of financial crises (Mandel 1977). This suggests that the regulatory framework provided by statutory town planning, rather than inhibiting the development of landed property as a relation of exchange (and investment), has encouraged this development. At the same time, speculation has presented new obstacles to particular uses of land, which is often interpreted as a 'crisis' in the field of town planning (Ravetz 1980).

Before the 1970s relations of landownership were largely ignored in studies of town planning. In more recent analyses, by contrast, landed property has been reintroduced into studies of this aspect of state activity (Lojkine 1976; Scott 1980). However, this is often done in a manner that ignores the historic importance of the state in constituting capitalist property relations. This omission is often the result of substituting an abstractly-conceived land market for the historical study of landownership. And, where the tradition of welfare economics is relied on, neoclassical assumptions are implicitly drawn into the analysis. By extension the state – in the first instance at least – is often portrayed as an agency external to an ahistorically-conceived land market. In studies as diverse as Scott (1980) and Lichfield and Darin-Drabkin (1980), this starting point provides the analytical basis for positing the inevitability of market externalities and breakdowns. These are conceived simultaneously as limitations to, and as the results of, the operation of the pricing system. Thus the analytical framework both derives and defends the need for state intervention. While neoclassical assumptions can provide the basis for provocative logical exercises, they are not substitutes for the concrete analysis of landed property.

In this essay I shall focus on the historical development of landownership in England in order to show that the state – through the agency of land law – has long been involved in shaping the development of landed property as a use value and as a basis for the appropriation of rent. As such, it has played a central role in subjecting this type of property to the commodity form. The key proposition is that the state, far from being opposed to the market, is pivotal to both the creation and the reproduction of property rights in land, which always have a specific form. In the following section I examine the role of the state in constituting rights in land, through law, as potential objects of exchange. Then, in the subsequent section, I elaborate on Marx's analysis of the contradictory nature of landed property in Victorian England, particularly the barriers posed by the estate system to development of land for specific uses. Finally I consider the way in which elements of modern town planning in Britain and the modern land market were jointly formed in response to the contradictions of landownership under the nineteenth-century system of great estates. I shall show that town planning, rather than simply reacting to market breakdowns, was part and parcel of the process of twentieth-century market formation.

The Early Market: The State and the Estate

Analysis of land markets must begin with the idea that land itself cannot be a commodity. What is bought and sold are legally-defined rights to spaces on the earth whose boundaries are marked by social institutions. Technically speaking, there is no land market, only a market (or more precisely, markets) in legally constituted rights in land. Under modern land law it is these rights, defined as rights both to use and to derive a revenue from land, which assume a commodity form.

Marx's analysis of capitalist relations of exchange provides a useful starting point for considering the development of property law. In his theory of value exchange involves a reduction of qualititatively different use values (or products of concrete labour) into quantitatively commensurate 'equivalents' (Marx 1967). The process of abstraction which he was struggling to reveal has to be understood as a very real process (see Elson 1979). The money commodity is one of the main results of this process (Weeks 1981). Its judicial dimension, on the other hand, is evident in legal norms that abstract from 'all and any specifics of time, place, parties and goods' (Kay and Mott 1982: 4). Under English land law, which is the special concern in this essay, the genesis of the estate in fee simple demonstrates abstraction. At common law, fee simple rights confer the greatest possible powers in land. Such rights are defined as both indestructible and unconditional (hence the term 'fee *simple*'). Both of these features have entered into the historical development of a freely-alienable right to landed property.

It must be emphasised, however, that these rights provided a necessary but not a sufficient condition for the emergence of a purely capitalist form of landed property. In Victorian England, the dominant form of landownership was the system of great estates, or the 'large landed property' to which Marx referred. On the one hand, the formation of the estate system presupposed the legal right to buy and sell land. On the other, the concentrated pattern of landownership on which it rested was itself a limitation to the full subordination of landed property to capital. Let us examine these arguments.

As late as the 1870s land in England was still being purchased for political as well as more purely economic ends – that is, it remained the perquisite of a social class. In a recent study, Massey and Catalano (1978) have rightly argued that this aspect of landownership prevented it from being a fully capitalist form of property. But the authors also point out, with equal merit, that the system of great estates was by no means a feudal legacy either. Quite to the contrary, the growth and consolidation of the estate system presupposed a history of legal developments that transformed precapitalist relations. Prominent examples are the seventeenth-century laws to bar entails and the eighteenth-century enclosures.

The law of entail was developed in the medieval period to help ensure the continuity of landed families. Through this law the owner of an estate in fee simple was able to create a 'perpetuity', a right in land that permanently barred its owner from selling it. An entailed estate was usually one that descended from father to eldest son, with the son being legally deemed the 'life tenant'. In having control of the freehold rather than the fee to the land (which remained with the father), he was prevented from alienating it for a period extending beyond his lifetime. Additional legal devices ensured that any land which he leased for inappropriate periods (or fraudulently sold as a fee simple estate) could be recovered when a new heir gained possession of the estate.

It is easy to see that the law of entail, if left unchecked, could have preserved late feudal patterns of landownership. But this did not happen. In the seventeenth century the development of the common law gave rise to modern laws against perpetuities. These laws were used to bar entails. As a consequence they extended the possibilities for a new class of landowners to gain access to landed property by means of the market.

Over the course of the late seventeenth and early eighteenth centuries there was a wave of land market transactions. This market activity was abetted by both the abolition of perpetuities and the development of new forms of taxation to replace the range of feudal incidences which were levied on landed property until the mid-seventeenth century. The purchasers of this period were generally 'not seeking good speculations, but well-tenanted estates that would yield a regular income with a minimum of trouble' (Habakkuk 1940; 14). With a regular income from ground rents in hand, landowners cast their gaze to the state and the benefits it might offer. The Acts of Enclosure, which Marx (1967) cites as a major form of primitive accumulation, were of special importance.

The Enclosures did much more than reallocate property rights; they redefined the nature of property rights themselves. Overlapping and inalienable rights to millions of acres of waste land were converted into fee simple estates. As a consequence the juridical basis was laid for the more extended development of a market in landed property.

Now the buying and selling of estates does not in itself imply that landed property is fully formed along capitalist lines. What is crucial is that the yield on investments in land be comparable with the yield on other investments. By the last quarter of the nineteenth century this was beginning to happen. Landed property was beginning to be treated as a relation of pure quantity, with the result that technical writers of the period were able to observe an inverse relationship between the price of land and yields on government-backed securities (Thompson 1957).

The statutes which facilitated the extension of this set of interlocking markets are as interesting as the market phenomenon itself. Many government interventions were involved, but the Settled Land Act of 1882 was crucial. By eradicating the effects of the legal encumbrances that had grown up around fee simple rights in land over the course of two centuries, this act served to simplify titles in land. In turn the property reforms of the seventeenth century were extended, and titles in land were made more comparable with personal forms of property (e.g. stocks and bonds). That this was the intended effect of the 1882 legislation is indicated in the debates which preceded its passage. These debates saw legislators calling for a statute which would 'either reocognise estates in fee simple, or give to the holder of land the same power of disposition which the holder of [a financial] stock now enjoys . . .' (quoted in Holdsworth 1946: 117). The Settled Land Act thereby served as a major basis for the 1925 Law of Property Act in which the fee simple rights of all individual landowners were formally recognised by the state.

The 1882 Settled Land Act shows the limited development of the nineteenth-century capitalist land market, which required encouragement through the active intervention of the state. The reasons lie above all in the nature of landownership within the great estates. By the mid-nineteenth century one of its most obvious characteristics was concentration. During the post-1846 'golden age of laissez-faire' (see Hobsbawm 1968), this resulted in Liberal agitation against the so-called land monopoly, together with modern Britain's first comprehensive survey of landownership. Published in the 1870s (and appropriately titled the *New Domesday Book*) it revealed that 'not more than 4000 persons' (Brodrick 1881: 165) owned nearly 19 million acres, or approximately 60 per cent of the land mass of England and Wales. Concentration of ownership was shown to be even more pronounced for Great Britain as a whole.

Barriers to Development Under the Estate System

The history of the great estates supports the proposition that law is central to the development of alienable property rights. This by no means implies, however, that 'the State [has] increasingly acted in one direction . . . changing definitions and practices associated with property so as to cumulatively establish, regulate and reproduce . . . interlocking markets' (Corrigan 1980: 41). The second proposition belies major discontinuities in the history of property law. If a statute such as the Settled Land Act (1882) is examined in isolation, then it is possible to see it as a mere link in a chain extending from the seventeenth century law against perpetuities to the 1925 Law of Property Act. Each of these interventions facilitated the transformation of property rights into marketable commodities. But when this legislation is seen in relation to eighteenth and nineteenth century laws designed to protect the great estates, the role of the state, at least as it related to landed property, is cast in a new light.

The presumption of linear development in property rights is common to modern legal texts and to many works of history. Both 'tend to portray the law of real property as a body of law which has zealously protected the power of free alienation of land, and the rule against perpetuities . . . as an effective curb against attempts to destroy this power' (Simpson 1961: 224). The assumption is shared by Scott in his *Urban*

Land Nexus and the State (1980), an important analysis of present-day landed property. He begins his anslysis with the neo-Ricardian proposition that 'the market' is always the mechanism through which 'the central class conflicts of capitalist society are registered' (1980: 16). In other words, for Scott the market is fully formed from the outset, even though his analysis takes in developments reaching back into the nineteenth century. Urban planning is portrayed as having grown up in response to the negative effects of the market, even though it serves to make those effects more pervasive.

The important example of Victorian railway construction, however, shows that state intervention was required because of the limited development of land rights as objects of exchange. Throughout the mid-nineteenth century, planning for railways involved the extensive use of compulsory purchase legislation to facilitate release of lands from the great estates. This type of intervention, besides expropriating tens of thousands of acres from the great estates, made all negotiations between railway companies and landowners subject to the potential force of the state. This, along with other early forms of state intervention, must be set within the context of laws which severely restricted the sale and outright purchase of landed property in the period.

Next to the House of Lords the Laws of Settlement provided the second main institution of the English 'landed class' (Thompson 1963: 64). These laws were extensively employed in the eighteenth and nineteenth centuries for the 'protection and consolidation of large landed property against the disintegrating pressures and vagaries of the capitalist market' (Anderson 1974: 56). The strict settlement differed from the medieval law of entail in that it did not create a perpetuity. Instead it gave rise to 'a life estate followed by a contingent remainder'. By this conveyancing device the owner of a landed estate could prevent his son from gaining possession of the fee simple, but he could not exercise similar powers over his grandson. In turn, this heir potentially stood to come into possession of an estate that would be his to sell or retain as he wished. Social and economic pressures usually forced him to resettle the estate, however. The result was that while strict settlements remained within the common law, they none the less circumvented its intended effects. Once settled, 'an estate tended thereafter to be bound by a chain of settlement and resettlement approaching perpetuity – son succeeding father generation after generation, each limited in his powers of alienation' (Spring 1977: 41).

The paradox of the Law of Settlement was that it functioned to maintain the integrity of the landed estates by reducing the potential size of the land market through which these estates themselves had been formed. In 'precluding land from being sold [*strict* settlements functioned as] real and effective instruments for the purpose intended: to keep the land together in large masses' (Mill 1948: 468). By the mid-nineteenth century their restrictive influence on the land market can partially be guaged by the extent of settled land in Britain. Contemporaries usually estimated that as much as 75 per cent of the land surface of the United Kingdom was affected by the laws of settlement (Offer 1980). The proportion of settled land in England and Wales was probably lower. But contemporary research suggested that 'there is no rashness in concluding that a much larger area is under settlement than at the free disposal of the individual landlords' (Brodrick 1881: 100). Brodrick's assessment of the situation has been partially corroborated by later research into the individual settlement agreements of the period (Thompson 1963).

The Laws of Settlement did not by any means create an absolute barrier to the sale or development of land through leasehold arrangements. By the mid-eighteenth century settlements often included clauses which allowed a certain amount of both. But where this settlement agreement did not permit leasing or sale the land could – pending an Act of Parliament – remain tied up for decades. The average period over which an estate owner remained bound by a particular settlement agreement was 35 years (Cooper 1976), and it was possible for the constraints of a settlement to apply for up to 70 years (Brodrick 1881). This meant that the development of land as an element of production in both agricultural and urban sectors could be inhibited.

In the rural sphere it was possible for a settlement document to prevent the sale of parts of an estate in order that investment capital be raised for the more intensive development of the land that remained. In the process the Laws of Settlement functioned to perpetuate the triadic system (landlord, farmer, agricultural proletariat) in which the tenancy relationship prevailed. And until the last quarter of the nineteenth century the leasing terms encouraged under the system were such that capitalist tenant farmers had no incentive to undertake extensive fixed capital investments. 'Is it possible,' asked Shaw-Lefevre, 'to conceive a system better calculated to prevent capital finding its way to the land?' (1879: 108).

Settlement agreements could also limit the development of land in the urban sphere. This was the case even when the agreement had been drafted as a guide to the positive development of estate lands. As one prominent historian of English land law has stated,

> even if a landowner wished to insert all the powers necessary to enable the tenant for life to manage and develop his estate, he might, in the age of rapid change which came in the nineteenth century, omit to insert the right powers. He could not be expected to foresee that the growth of a neighbouring town might, fifty years hence, make part of his estate valuable building land. Nor could he foresee that the discovery of minerals might convert an agricultural estate into an industrial centre. (Holdsworth 1926: 211)

The settlement agreement, once again, was not an absolute constraint on development. It could be changed through a Private Act of Parliament, and a large number of such acts were passed. But their costs, together with the incentive provided by the settlement agreement to leave undeveloped land to appreciate, meant that the Laws of Settlement probably substantially influenced the degree to which the estate system presented a barrier to urban growth.

The Laws of Settlement also presented governments of the mid-nineteenth century with a dilemma. These laws could be left intact, in which case their contradictory effects would intensify, or they could be abolished. Abolition had two advantages. First, it would increase the ability of individual landowners both to meet the threats of foreign agricultural competition and to accommodate the growing pressures of extended urban development. Second, it would remove the legal focus of the agitation against the landed class in general. The disadvantage was the great estates might disintegrate under the pressure of market forces.

During the 1850s legal and fiscal protections for the system of great estates were supported on political and ideological grounds. In 1856 Lord Palmerston wrote to his Attorney General, stating that

> I consider it essential to the proper workings of our constitution to preserve as far as possible the practice of hereditary succession to unbroken masses of landed property. That practice was much broken in upon by the application of succession duties to landed property; that measure was, however, deemed necessary, but I could not be party to any further inroad upon a principle which I consider of great political importance . . . (quoted in Select Committee on Small Holdings, 1889, Question 6011)

This position – and the support for the Laws of Settlement which followed from it – was undoubtedly shaped by the numerical dominance held by the landed interest in both Houses of Parliament down to the 1880s.

Yet the political power of the representatives of landed property was not so great as to prevent the expropriation of land for railway development. This occurred through the thousands of private Railway Acts passed between 1830 and 1870. But the costs of compensation payments to estate owners were enormous. Costs of this sort are appropriately regarded as the price that was paid for the ongoing protection of the estate system under Settlements.

With respect to agricultural and urban development, a series of partial reforms provided a way out of the dilemma presented by the Laws of Settlement. Urban development was to be facilitated by the 1856 Sales and Lease Act. It partially overrode the inflexibility of settlement agreements while also ensuring that the long-term interests of the estate would be protected. Under the Act life tenants could be released from settlement restrictions at the discretion of the Court of Chancery, which had jurisdiction over landed property. However, the Court allowed exemptions from settlement agreements only on condition that there was 'due Regard for the Interests of all Parties (present and future) entitled under the Settlement.' (Leases and Sales Act, 1856, 19 and 20, Vict., Cap. 120.) And before it could grant new leasing powers, it had to establish in each individual case that (i) 'on every such lease shall be reserved the best rent . . . either uniform or not, that can reasonably be obtained'; and that (ii) 'every such lease shall contain such convenants, conditions and stipulations as the Court shall deem expedient with reference to the special Circumstances of the Demise'. While facilitating land development provided the *raison d'être* for the 1856 Act, the policy of protecting the physical integrity of the landed estate still weighed heavy. Except in extraordinary cases, new powers of sale were to be limited to the outlying portions of a settled estate. The result was that the legislation did not lead to the general release of landed property from the estate system. Instead it merely extended the degree to which land was reshuffled within the system.

In effect the state was intervening to break individual settlements without going through the inconvenience and expense of a larger number of Parliamentary Acts. In the process the landed estate was transformed into an object of administration. The major problem with these reforms, however, was that the agencies which were given the responsibility for overseeing their implementation did not generally function efficiently as administrative (or quasi-judicial) agencies. Thus land development continued to suffer under the Law of Settlements.

Beyond this, it led to the charge that the partial reform served only to substitute 'the real owners of property by a government department, a state inspector, or a judicial tribunal' (Shaw-Lefevre 1879: 119). Between 1846 and 1882, such charges were made in conjunction with demands for 'free trade in land'. This was the campaign call of the law reform movement. It reflected the view that abolition of the Laws of Settlement would make landed property more accessible through the market, as well as hasten the decline of the landed class.

In summary, the state took the first step in creating a market in landed property through its interventions to facilitate the development of the estate system. This system, however, presented barriers to the use of land as an element of production and reproduction. By the end of the nineteenth century the state had created a complex situation in which land was partly open and partly blocked to capital. The process of further opening the sphere of landed property to capital presented dangers to society along with the opportunities.

Planning: From the Estate to the State

By the mid-1880s the state confronted a series of problems arising from two aspects of landed property: the market in land remained restricted even after the Settled Land Act of 1882; at the same time, the estates did protect land from the socially destructive uses that unfettered investment might bring. Although legislation was not coherently conceived in this way, laws increasing the access of capital to land, and those establishing modern town planning, were simultaneously and mutually reinforcing.

The estate system dominated the overall process of urban development in centres such as London until the last third of the nineteenth century, and even then the developmental effects of this system remained pervasive. The positive implications of this history can be seen today in inner areas of the city such as Chelsea and Belgravia. Of course, urban development within the landed estates which dominated these areas benefited from proximity to London's West End. But as the work of Cannadine (1980) and others (Olsen 1963; Thompson 1974) indicates, the aristocratic pattern of estate management was crucial to the comprehensive ordering of development within these areas.

In the first instance, this form of estate management was orientated to the preservation and enhancement of the landed family's long-term social and financial position. To these ends the leasehold system was used in order to maximise the potential for the unified management and control of estate lands. Secondly, and as a corollary, the overseers of large landed estates often had 'neither the desire to maximise profits, nor the immediate financial concern, as a major preoccupation' (Cannadine 1980: 392). And even if they did, settlement agreements served as legal vehicles for maintaining the traditional concerns of landed families. These agreements endowed the landed estates with a semi-corporate quality. One of their main purposes, writes Thompson (1963: 66), 'was to protect the estate from the worst depredations of spendthrift heads of family. The head for the time being was only a life tenant . . . and therefore unable to do any lasting damage to the inheritance of generations to come.' Powers of sale were limited, and leasing powers, while more extensive, were usually complemented by specific provisions designed to maintain the long-term reversionary value of estate lands.

Over the course of the nineteenth century, the concern to maintain the long-term value of landed property evolved into the short-term practice of denying the working-class access to large areas of land in and around metropolitan centres such as London. The legacy of this practice can still be seen in the many parks in London which remain barricaded by iron gates. There were a range of exclusionary devices. For example, small speculative builders (who were forced to 'build down to cost') could be kept at a distance by conditions in ground leases stipulating the need for the expenditure of large amounts of capital per residential unit. Because the market for more expensive housing was limited, however, vacant units and land were often the result. Ironically, the more such devices were used, the greater was the likelihood that working-class overcrowding would be exacerbated on the land not subject to them. The intensified process of slum formation in turn encouraged the use of restrictive practices to protect the long-term value of estate lands. This pattern of development underlies the view that the great estate owners attempted to 'deny that the "lower orders", retail trade or public transport existed' (Cannadine 1980: 220).

Victorian slum conditions, which have been extensively analysed (see Gauldie 1976; Jones 1971; Wohl 1977), were the underside of the ordered development which usually followed from the use of planning practices on the great estates. It should be emphasised that these practices were carried out under the Laws of Settlement and under statutes, such as the 1856 Sales and Leases Act, discussed above. Their negative consequences, exacerbated by government-sponsored road, railway, and slum 'clearances' (Jones 1971), inspired public 'planning schemes' introduced under the Housing and Town Planning Act of 1909. These schemes were developed as administrative mechanisms to secure the planned release of land for the housing of the working class in suburban areas. The Act gave administrators regulatory powers designed to match those held by the managers of the great estates.

While the background to statutory town planning can be found in compulsory purchase, it is distinct in a number of crucial ways. First of all, it was formally established in Britain during a period when the political and legal supports of the dominant form of nineteenth-century landed property were being removed. One of the main implications was that legislators could assume that individual landowners had the power to dispose freely of their property, even if they did not exercise it. A second implication was that the planning legislation emphasised regulation instead of expropriation. The political significance of the 1909 Planning Act was that it established that losses suffered by landowners as a result of the imposition of land-use restrictions were not compensatable.

The passage of the 1909 Act, as William Ashworth (1954) says, was the first step in the constitution of the quasi-judicial framework within which modern town planning takes place in Britain today. Ashworth's analysis exposes an important aspect of town planning: public land-use controls were made increasingly comprehensive from the 1860s onwards. His account of the early 'exemplars' to the town planning movement also reveals some of the visions that inspired it. However, he fails to examine the explicit understanding of the nineteenth-century urban reformers that the suburban solution to the housing problem presupposed the extension of state powers over landed property. He does not tell us, for example, that Raymond Unwin, in addition to being one of the principal architect-planners of Hampstead Garden Suburb, also held the view that

> the improvement of the towns is checked by the extravagant prices demanded, and which must be paid [for land]. You have got a ring, a dense ring, of impenetrable greed compressing the town and crushing it in until at its very heart you get a hard slum. We want to shatter that ring, so that the town shall have lungs and expand . . . (Unwin 1912: 70)

This view stemmed from the widely-held perception that the great estates presented an extensive barrier to the suburban development of working-class housing. It was shared by prominent politicians, urban reformers and Royal Commissioners from the mid-1880s onwards.

Thus contemporaries were aware that, as I argued above, the Settled Lands Act of 1882 was not sufficient for the break-up of the great estates, however much it facilitated their purchase and sale. The break-up of the estates began in the three-year period before the First World War (Thompson 1963). Before that, there was a major struggle over proposals to use taxation as a new form of intervention against landed property. The contents of these struggles, and of their relation to town planning, were part of a larger project to reduce landed property to 'a more fluid form of capital' (Lord Haldane, quoted in Emy 1971: 55).

The 'land clauses' in the so-called People's Budget of 1909–10 directly precipitated a constitutional crisis (Mallet 1913; Murray 1980). The crisis was over the right of the House of Lords to veto government legislation (and thereby remain a political bastion of landed power). The political dimensions of the Finance Bill were sagaciously described by two contemporary commentators:

> Muncipal rating of unimproved land values, which the great cities have asked for again and again, is at present impossible owing to the power of the House of Lords . . . Until its power of obstruction is taken away, the House of peers will stand between the Chancellor or the Exchequer and the land values of England (Chomley and Outhwaite 1909: 148).

The authors proposed that the Budget, over which the Commons nominally had control, was the key to the situation. They advocated that the Budget include taxation of land values, even though valuation clauses necessary for this form of intervention were likely to lead to its rejection by the Lords.

All this indeed came to pass. Finance Bills containing valuation proposals were twice vetoed by the Lords, in 1909 and 1910. Each veto precipitated a general election. After the second Liberal victory, the Parliament Act of 1911 at once subordinated the Lords to the Commons and initiated national site valuation. The valuation clauses were designed to provide the information necessary for the taxation of the site value of landed property. Under this form of taxation the 'capital value' of land (i.e., the capitalised value of rents derived from its most profitable use), as opposed to current rental income (in whatever use), was to be the basis for taxation. It was a fiscal reform intended to force landowners to respond to market pressures. The assumption was that an acre of land developed for 20 houses not only yielded a higher ground rent than neighbouring land held vacant, but also that it yielded greater benefits to the community. It follows that both plots would be assessed at the same capital value and taxed at the same rate. In other words, advocates of valuation expected that if the short-term demands of the market were met with a corresponding supply of land, then benefits to the community and to individual proprietors would be in harmony.

Opponents of site value taxation argued that it would actually increase overcrowding through its effects on the supply and price of suburban land. The various proponents of this argument all agreed that, lessees permitting, site value taxation would force those owners who were not maximising short-term rental revenues to release land for more intensive development. However, they argued that an increased flow of land onto the market would merely reproduce the emerging problems of suburban overcrowding throughout the urban fringe. Thus the problems of suburban monotony and environmental deterioration would be extended on a massive scale.

Advocates of fiscal reform did not dispute these arguments, but met them with proposals for town planning in conjunction with site value taxation. John Thompson, one of Ashworth's 'exemplars' to the town planning movement and Chairman of the National Housing Reform Council, argued forcefully that

> Any system of putting a tax on land values must be accompanied by the power of town planning in order to prevent a lot of jerry-building and land being forced into the market in the wrong places at the wrong times. It is one of the strongest objections to a tax on land values, that it would tend to push land into the market and to get a lot of wretched suburbs created as we now have. (Parliamentary Papers, vol. IV, 1906, Q 2669)

Then, more explicitly, he stated before the same Select Committee (Q 2674) that

> before you have a taxing of land values you must have your town planning and your town and village development schemes in order to have a really successful working.

Thompson, like others who wished to open landed property to urban expansion, understood that stronger market forces had to be accompanied by town planning.

Economist A.C. Pigou (1909: 17) lent his expertise to the argument, writing that the best uses of land for the community did not in the aggregate correspond to those dictated by market values. For example, estate gardens had indirect hygienic advantages to the community, which would not be reflected in prospective rents or capital values. Pigou proposed that if some of the advantages bequeathed by the great estates were to be preserved, the state would have to both plan for the retention of certain types of land use (directly at public expense) and develop its ability to impose discriminatory taxation. Others called for comprehensive regulation of land uses to prevent intensification of urban problems associated with the atrophy of the estate system and to ensure that the decline would be smooth. Indeed, it was claimed that planning was a necessary supplement to taxation even to achieve release of land to the market. An eminent contemporary wrote, 'remove by town planning legislation the fear of the jerry-builder, and at once large blocks of land will come into the housing market all over the country' (Nettlefold 1910: 70).

The merits of planning were understood as the maximisation of aggregate rents from urban land. Maximal aggregate rents presupposed both the development of inexpensive housing and the assumption by public agencies of the role traditionally played by larger estate agents: the guarding of long-term development and reversionary values through comprehensive measures to regulate land uses. Under the estate system, such control had been exercised through restrictive convenants, building agreements, and the right to withhold land from uses for short-term profit. The 1909 Housing and Town Planning Act, by contrast, established planning schemes based on public land-use regulations. Land became the object of public administration.

Land-use control, together with extension of compulsory purchase for planning purposes, was the main element of the town planning framework set up in 1909. It could be – and eventually was – used to subordinate the interests of individual landowners to those of ordered urban development. But as one forward-looking Conservative spokesman observed (*Hansard*, April 12, 1909, col. 855):

> The principle of town planning will extend the market for building. Instead of being almost forced to sell small parcels of land upon which houses are packed at immense inconvenience . . . the tendency

of this legislation ought to be to extend the zones in the suburbs, to spread the building operations, and therefore to bring much more land into profitable occupation.

Despite such insights at the time, there was intense struggle over the Housing and Town Planning Act of 1909. The 'frontal assault' on the land monopoly (Thompson 1963: 321), far from being stopped by the veto of the House of Lords, was increased (Emy 1971). It precipitated the break-up of the estate system, which began in the years between 1911 and 1914. Then in the interwar periods, as taxation continued to erode the estates, town planning developed in tandem, as the framework established in 1909 was extended through Planning Acts of 1919, 1925 and 1932. This legislation, plus the 1925 Law of Property Act (see Ball 1983), facilitated waves of suburban development. In the process, landed property was transformed into a more accessible object of speculative investment.

Conclusion

The state has consistently been central in constituting rights in land and in responding to contradictions between land and capital. The state, therefore, is not external to the sphere of landed property. It has not simply 'intervened' in an autonomous sphere; still less can its role be understood in a linear way. As late as the nineteenth century landed property was partly capitalist, partly not. Its relation to the larger capitalist society was continually reconstituted through changes in the laws governing sale and use of the great estates.

The land market did not emerge fully-formed with early capitalist property relations. Instead primitive accumulation through enclosures gave rise to a form of landed property which confined marketable rights in land within elaborate legal constraints. The laws that established and governed the reproduction of the system of great estates, simultaneously created and restricted the possibilities to buy and sell land. As a result, the ability of capital to develop the land was often severely circumscribed. This had both positive and negative effects. Positively, private estate planning was guided by long-term concerns for the whole of the great estate, for which few costs could be made external. Negatively, the long-term concerns of landowners inhibited the development of land as an element of production, especially in agriculture, and of social reproduction, especially in working-class housing. The state, through a complex but inter-related set of laws, reconstituted the market, and took the planning role of the estate into the public realm.

The British state, having created the great estates in an earlier period, began in the early twentieth century to break them up. It replaced the estate system with an integrated extension both of private rights in landed property and of public administration. From the Settled Land Act of 1882 to the Housing and Town Planning Act of 1909, government responses to the contradictions of landed property involved the double development of land as an object of administration and of private law. The state became the point of unity between planning and the market. Yet the contradictions have not been overcome. They have been transformed through the subordination of landed property to capital.

The state, from which property rights and planning originate, has created new contradictions by extending possibilities for speculative investment in land. The effective break-up of the great estates in the interwar years led to fears of speculation which were expressed in the Uthwatt Report of the early 1940s. By the 1970s these fears were being realised.

Acknowledgements

I would like to thank Michael Ball, Michael Edwards, and especially Harriet Friedmann, for critical judgement and support. Conversations with Geoffrey Kay provided important background for the arguments on law and value.

References

Anderson, P. (1974) *Lineages of the Absolutist State*, New Left Books, London

Ashworth, W. (1954) *The Genesis of Modern British Town Planning*, Routledge and Kegan Paul, London

Ball, M. (1983) *Housing Policy and Economic Power: the Political Economy of Owner Occupation*, Methuen, London

Brodrick G.C. (1881) *English Lords and English Landlords*, Cobden Club, London

Cannadine, D. (1980) *Lords and Landlords, the Aristocracy and the Towns 1774-1967*, Leicester University Press, Leicester

Chomley, C.H. and R.L. Outhwaite (1909) *The Essential Reform: Land Value Taxation in Theory and Practice*, Sidgwick and Jackson, London

Cooper, J.P. (1976) 'Patterns of Inheritance and Settlement' in J.P. Goody, J. Thirsk and E.P. Thompson (eds.), *Family and Inheritance, Rural Society in Western Europe*, Cambridge University Press, Cambridge, 192–328

Corrigan, P. (1980) 'Towards a History of State Formation in Early Modern England' in Corrigan (ed.), *Capitalism, State Formation and Marxist Theory*, Quartet Books

Elson, D. (ed.) (1979) *Value, the Representation of Labour in Capitalism*, CSE Books, London
Emy, H.V. (1971) 'The Land Campaign: Lloyd George as a Social Reformer, 1909-1914' in A.J.P. Taylor (ed.), *Lloyd George Twelve Essays*, Hamish Hamilton, London, 35-68
Gauldie, E. (1976) *Cruel Habitations, A History of Working-class Housing 1780-1918*, Allen and Unwin, London
Habukkuk, H.J. (1940) 'English Landownership, 1680-1840' in *Economic History Review*, vol. X, 2-17
Harvey, D. (1982) *The Limits to Capital*, Basil Blackwell, Oxford
Hobsbawm, E. (1968) *Industry and Empire*, Pelican Books, London
Holdsworth, W.S. (1927) *A Historical Introduction to the Land Law*, The Clarendon Press, Oxford
—— (1949) *Essays in Law and History*, The Clarendon Press, Oxford
Jenkins, S. (1975) *Landlords to London*, Constable, London
Jones, G.S. (1971) *Outcast London*, The Clarendon Press, Oxford
Kay, G. and J. Mott (1982) *Political Order and the Law of Labour*, The Macmillan Press, London
Lichfield, N. and H. Darin-Drabkin (1980), *Land Policy and Planning*, George Allen and Unwin, London
Lojkine, J. (1976) 'Contributions to a Marxist Theory of Urbanization' in C.G. Pickvance (ed.), *Urban Sociology*, Tavistock Publications, London, 119-46
McMahon, M. (1982) *Town Planning and the Development of a Land Market in Britain, 1845 to 1910*, unpublished M.Phil Thesis, University College, London
Mallet, B. (1913) *British Budgets*, The Macmillan Press, London
Mandel, E. (1977) *The Second Slump*, New Left Books, London
Marx, K. (1967) *Capital, I*, International Publishers, New York
—— (1973) *Grundrisse*, Penguin Books, Harmondsworth
Massey, D. and A. Catalano (1978) *Capital and Land: Landownership by Capital in Great Britain*, Edward Arnold, London
Mill, J.S. (1948) *Principles of Political Economy*, W.J. Ashley (ed.), Longman Green and Co., London
Murray, B.K. (1980) *The People's Budget 1909-1910: Lloyd George and Liberal Politics*, The Clarendon Press, Oxford
Nettlefold, J.S. (1910) *Practical Housing*, Garden City Press, Letchworth
Offer, A. (1981) *Property and Politics, 1870-1914*, Cambridge University Press, Cambridge
Olsen, D.J. (1963) *Town Planning in London*, Yale University Press, New Haven
Pigou, A.C. (1909) *The Policy of Land Taxation*, Longman Green and Co., London
Ravetz, A. (1980) *Remaking Cities*, Croom Helm, London
Scott, A.J. (1980) *The Urban Land Nexus and the State*, Pion, London
Shaw-Lefevre, G. (1879) *Freedom of Land*, MacMillan and Co., London
—— (1893) *Agrarian Tenures*, Cassel and Co., London
Simpson, A.W.B. (1961) *An Introduction to the History of Land Law*, The Clarendon Press, Oxford
Spring, E. (1977) 'Lawyers and Land Reform in Nineteenth Century England' in *The American Journal of Legal History*, no. 21, 34-53
Thompson, F.M.L (1957) 'The Land Market in the Nineteenth Century' in *Oxford Economic Papers*, vol. IX, no. 3, 285-308
—— (1963) *English Landed Society in the Nineteenth Century*, Routledge and Kegan Paul, London
—— (1974) *Hampstead: Building a Borough 1650-1964*, Routledge and Kegan Paul, London

Unwin, R. (1912) *Nothing Gained by Overcrowding*, Garden City and Planning Association
Weeks, J. (1981) *Capital and Exploitation*, Princeton University Press, New Haven
Wohl, A.S. (1977) *The Eternal Slum*, London

Parliamentary Papers

PP. (1884–1885) XXX c. 4402: *Royal Commission on the Housing of the Working Classes*
PP. (1889) XII c. 313: *Select Committee on Small Holdings*
PP. (1941–42) IV c. 6386: *Expert Committee on Compensation and Betterment* (the Uthwatt Committee)

6 LAND, CAPITAL AND THE BRITISH COAL INDUSTRY PRIOR TO WORLD WAR II[1]

Ben Fine

For the purposes of analysis, the British coal industry in the century leading up to the Second World War is usually considered in terms of two periods, the one neatly separated from the other by the First World War. On the face of it, this appears to be justified by the contrasting fortunes of the industry over the two periods concerned. From 1850 to 1913 employment increased from 250,000 to over a million. Output increased from 60 million to 300 million tons with the proportion of exports rising from under ten to over thirty per cent and accounting for a tenth of the value of all British exports by the latter date.[2] By contrast the industry in the interwar period is scarcely ever seen to be free of either short- or long-term problems with special emphasis being placed upon unemployment, industrial strife, collapsed export markets and the failure to reconstruct the industry in almost every respect. The problems of the industry were already recognised at the end of the First World War and could be substantially documented in the Samuel Report (1926). The Reid Report of 1945 had very little extra to add in substance rather than detail given the intervening passage of twenty years. Nevertheless its conclusions were more palatable by that time and made more so no doubt by their conciliatory but disingenuous observation that:

> When we come, therefore, as we must, to point out the mistakes which were made in these early years of the coalmining industry, let us beware of merely being wise after the event, or of withholding the needs of praise due to a great race of men, employers, mining engineers, workmen and machinery makers alike. For whatever their faults, they were fit to rank with the greatest of Britain's industrial pioneers.

Output was never to recover to its pre-war level and export markets were lost permanently. Whatever the extent to which this can be explained by the decline of a staple industry in depression conditions, particularly in the thirties, it does not directly explain the dismal record

Table 6.1: Comparative Performance of British, Polish and Ruhr Coalfields

	Output per Manshift, cwt			Average
	Date Indicated	1936	% Increase	mine size '000 tons
Poland	23.4 (1927)	36.2	54	750 (1937)
Ruhr	18.6 (1925)	33.7	81	780 (1938)
Britain	20.6 (1927)	23.5	14	120 (1938)

of reorganisation that is indicated by comparison with European competitors (see Table 6.1). Apart from the failure to concentrate production on fewer, larger mines, Britain fell behind in mechanisation, with coal cutting, for example, being limited to 55 per cent of output in 1938 compared to 97 per cent in Germany, 99 per cent in Belgium and 88 per cent in France.[3] It is this failure of performance, particularly in comparison with the period prior to the First World War, that has been the concern of the economic historians of the industry.[4] Generally these historians go little further than pointing to the symptoms of failure, such as the poor record of mechanisation or amalgamation, or they rely upon the residual explanation of poor entrepreneurial (or worker) performance.[5] Where even this proves inadequate, it is always possible to appeal to supposedly exogenous factors such as the depressed state of demand or the influence of geological conditions.[6]

In this paper, we place attention upon the role played by landed property in the development of the British coal industry. We attempt to show that the industry's changing fortunes were in part a production of the impact of the private ownership of land. In doing so, we put to one side the significance of the capital-labour relationship by concentrating exclusively upon the capital-land relationship. This does not reflect a view that the role of land is more important, quite the opposite, although we would emphasise the interrelationship between the three classes involved.[7] Rather our intention is to stake a claim for the importance of examining the significance of landed property in capitalist industrial development. It is further to emphasise the necessity of understanding the specific historical circumstances in which landed property intervenes in the accumulation of capital. Attention cannot merely be confined to a treatment of rent as a distribution of surplus value. Nevertheless, despite our focus upon the relationship between capital and land, we do include a section considering the attitude of labour itself to the role of land in the development of the industry.

Coal Royalties

In Britain, the ownership of all minerals other than gold and silver was held by those who owned the surface of the land. This followed a judgement in the case of the Queen versus Northumberland in favour of the latter in 1568 (see Nef 1932). Only in 1938 were the coal royalties finally taken into state ownership in Britain. Certainly by the end of the nineteenth century, even though landowners may occasionally have undertaken to mine the coal that they owned, they generally did not do so and, when they did, usually also leased land to other mineowners.[8] It is also important to recognise that the coal royalty was very often not paid on a per tonnage basis even though statistics of royalties almost invariably give this impression by dividing total royalty by total tonnage. The leases governing royalties and their amounts were both complex and varied. Accordingly, the mines were generally owned and run separately from the ownership of land so that leases for coal royalties had to be negotiated between the two parties of mineowner and landowner. This situation differed sharply from that on the continent where coal royalties had in general been taken into state ownership around the turn of the nineteenth century.

In the 1880s, a fear became expressed that the private ownership of royalties could in part be responsible for raising the price of coal by the royalty levied, thereby damaging the industry's own export competitiveness and the competitiveness of industry in general due to the higher price of an essential (power) input. In conformity with the fashion of the time, a Royal Commission was set up to investigate the question. The Commission essentially dismissed the fear as groundless. It employed a theory, reflecting the date, by bridging Ricardo and marginalism, arguing that the system of landed property had minimal effects other than to allocate capital efficiently between mines with royalties squeezing out the differential surplus associated with superior geological or locational conditions. Whatever their differences Ricardo and Jevons could agree on this. In his work on the coal industry, Jevons (1865) could adopt the Ricardian notion of decline into a stationary state in which, for Britain, the extensive margin or corn for Ricardo had become displaced by the extensive margin of coal extraction, the product to be exported in trade for corn. At a more pragmatic level, the Commission suggested that the industry's successful expansion spoke for itself and was evidence that the royalty system could not be an impediment to its development. At this stage, the Commission conveniently forgot that the same success could not be adduced in the case of iron ore extraction (also within its remit) which was already experiencing the difficulties that were later to plague the coal industry. The industry was, however, very much smaller. There were also problems with Ireland for coalmining where landownership was much more dispersed. Nevertheless, the Commission undertook a detailed empirical investigation of the royalty system both for Britain and the rest of the world.

It discovered that the reason why the royalties had been nationalised on the Continent was that the ownership of landed property had been so fragmented that mining could hardly have begun without the consolidation of the right to mine coal independently of surface ownership. The reasons for this were made clear by representatives from various countries who were questioned by the Commission. The prevailing patterns of landownership were so subdivided that, for a reasonably sized mine to be established, terms would have to be arranged with many separate landowners. Thus,

> it is unanimously admitted that the fertile results follow the absolute distinction that exists between surface property and the working right of mines. In a country where property is so minutely subdivided as it is in France, the reasonable and active working of mines would be impossible on any other system (than state ownership).

For Germany 'in many industrial districts of the country the ownership of the surface is so divided that it would be impossible to carry on deep mining under any other principles' and

> Besides, in many cases a strata of minerals extends underneath the property of several landowners, and it would be almost impossible to work different mines scattered on the larger or smaller plots belonging to different owners, and to this would come the additional difficulty of arriving at an agreement between the owners with regard to the working of mines . . . it has been arranged in Austria-Hungary to make them (minerals) entirely independent from the landowner . . . This system has specially promoted the establishment of mines.

The same story is also told for Spain, Portugal, Italy, and Luxembourg; landownership is so fragmented that minerals had been taken into state ownership to promote the development of mining. Without this, capital could not flow freely onto and between lands. By contrast, for Great

Britain this problem scarcely seems to have been considered by the Commission who simply observe that 'where a large mineral field is the property of one individual no difficulty arises in respect to its full development.'

The reason for this is not difficult to discern. In Britain the pattern of landownership was not fragmented; ownership was highly concentrated and much the same was true of royalty ownership. Rather than small landowners obstructing mining through the charges that would be made for the small quantities of coal that they owned, it was more a case of large landowners encouraging a number of mineowners to extract as much coal as possible, and this explains the occurrence of fixed rents to be paid irrespective of the quantity of coal removed but against which royalties were set.

It is this pattern of large landed property that made nationalisation of the royalties unnecessary in Britain. The same is also true of those countries such as India, Australia, America, and Canada, where English law prevailed but large concessions and often the coincidence of land- and mine-ownership had been created. In other words, it is not the state or private ownership of royalties as such that was important but the extent to which large enough coalholdings could be leased to form mines of adequate size. Indeed, in many of the European countries, mineral rights were no sooner taken into state ownership than they were sold as concessions to private individuals. These concessions could then be traded and amalgamated. Further regulation of this to prevent concessionaires obstructing the development of mining through outrageous charges does not seem to have been a major problem. Where it was, state regulation was still possible and this could also be used to guard against monopoly of supply. Thus 'the tendency in Belgium is towards the amalgamation of several neighbouring concessions, these being often of small extent,'

> It is a common practice for concessionaires to sell or let their concessions to companies who undertake the working of the mines . . . There is absolutely no guard against companies enlarging their holdings through purchase of other undertakings: in fact this is proceeding very rapidly in all the German coal fields without check.

'In the north of France at least, concessions were commonly united and are generally worked by companies' and 'in Austria-Hungary concessions can be and are freely sold'.

By contrast, in Britain, the situation was different. There was large landed property[9] but this enabled landowners to encourage and provide for one or more mineowners to employ their capital upon the land. Accordingly it is not state ownership of land as such that was necessary for mining development on the continent, but the creation of power over sufficiently large mining leases. In fact, the European states often granted leases which then became the subject of private exchange, rendering state ownership a mere formality. In these terms, the system of landed property and the pattern of ownership in Britain can be seen to have been conducive to the early development of the industry with the estate owners having an incentive in estate development to facilitate the extraction of coal (and royalties) by many capitalists.

The Commission also discovered that, despite its generally favourable impression, problems were arising at the boundaries between properties. These problems were most notable in transporting coal over adjacent property, whether under or over ground, but they also concerned the extension of leases. It was generally thought that such difficulties could be overcome by sensible negotiation, but in retrospect they are symptomatic of the related developments. With the early extraction of the most easily won coal, pits would become deeper and wider to justify the extra fixed capital expenditure in setting up a pit. Necessarily, the land requirements of a mineowner would expand and tend to exceed the boundaries encompassed by a single estate. In conjunction with this then, the capital required by a mineowner would tend to expand to pay advances in royalties as well as to make fixed expenditures for shafts etc. This perhaps explains the declining participation of landowners in the extraction of their own minerals given the other avenues to which their financial resources could be devoted.

Given rapidly expanding industry, it is clear that these problems, or more exactly the conditions for them and hence other problems as well, would themselves quickly intensify. This is exactly what occurred and it is the speed with which they matured which is perhaps surprising. Even before the First World War and little more than a decade after the Royal Commission a serious and general difficulty concerning liability for subsidence arose.[10] Essentially, the pre-existing law assumed that subsidence would only occur immediately above a tunnel. Mines of greater depth proved otherwise by damaging neighbouring properties. It took ten years to sort this out between the three major interested parties, the mine, royalty and railway owners. By the end of the First World War, even respectable public opinion had been reversed. Indeed, every relevant major state body in the interwar period favoured nationalisation of the royalties from the Sankey Commission of 1919 onwards. Even the mineowners concurred at this early stage, although their opinion fluctuated according to political expedience and changing legislation.

More favourable legislation to the mineowners over access etc. was passed in 1923 and amended in 1926, but even on its own terms it was ultimately to prove inadequate in the courts in the 1930s. This in conjunction with the fear of the working class in the political climate of the early twenties led to a turnabout of mineowner opinion by the time of the Samuel Report of 1925. As the Coal Association (1920) argues,

> The government proposes to acquire mineral rights but not mines and to pay compensation. The Miners' Federation proposed to pay no compensation and also to go onto mines, land etc. . . . this makes the duty of Parliament all the more imperative to reject 'the thin end of the wedge' viz. the Nationalisation of Royalties . . . What justification is there for not nationalising the other substances or minerals . . . or other industries for that matter?

This treatment of the mineowners' opinion is a more or less accurate assessment of the balance of views of a variety of extremely individualistic entrepreneurs.

What exactly was the nature of the problems involved and why were they so drastic that even members of the ruling classes were proposing nationalisation as a solution? The Scott Committee (1919) indicated 14 problems, predominantly concerning itself with difficulties associated with the division between property rights. As it had taken ten years to negotiate one of these problems between interested parties, a certain impatience on the part of those concerned may be appreciated even if in principle the other problems could have been achieved through negotiation in the very long run. The negotiations over subsidence had taken so long and were so delicately balanced that it was simply insisted that no amendments could be made to the enabling legislation of 1923 as it passed through Parliament. Another publicised problem concerned the leaving of coal barriers to demarcate underground lines of leases held by mineowners. The estimate of accumulated lost coal was 4,000 million tons. Another concerned the 'externalities' governing the draining of water for which efficient organisation of dams and pumping was impeded by the need for drainage systems to conform to the dictates of property rights.[11]

Despite the early recognition of these and other problems, the arguments for nationalisation and against the private ownership of royalties concerned more general problems of reorganising the industry on a 'national' basis to take account of lay-out, all forms of mechanisation and the need for large-scale mining. For this, the ownership of royalties was now seen to be too fragmented and dispersed. The Samuel Report made much of the fact that five leases were required on average by each mineowner[12] and saw this as a major impediment to the industry. The same viewpoint is expressed consistently from 1930 to 1946 by Sir Ernest Gowers who was charged with reorganising the industry during the thirties. In short, it appears that the impediments to the establishment of a coal industry on the continent created by the system of landed property were themselves the product of the industry's own development in the case of Britain. This was so much that the political pressure for nationalisation of the royalties from within ruling circles was substantial even if political expediency delayed implementation up to 1938.[13]

Significantly the arguments for nationalisation of the royalties in the interwar period concerned practical matters; which sensible individuals would recognise if they were not infatuated with the idea of the perfect workings of the market. Equally significantly the opponents of nationalisation recognised the difficulties and proposed schemes of unification of the royalties which would fall short of state ownership. Moreover they could bring forward precisely the same theoretical arguments as the earliest Royal Commission. The system of landed property made no difference other than to determine the recipients of royalties which were fixed by geological differences. Consequently, either royalty owners would be fairly compensated, in which case the state would constitute an unnecessary intermediary (with costs), or they would be unfairly compensated and the royalties would simply fall to somebody else with no other effect.

To our knowledge, the proponents of nationalisation never answered these theoretical arguments nor put forward any of their own. The practical arguments appeared to suffice. This is quite common in questions concerning land tenure. Economic theory, particularly but not exclusively neo-classical, tends to treat the system of landed property as irrelevant, since land is treated as a thing defined by physical properties rather than as always being associated with the social relations governing access to it. Accordingly, when the system of landed property is conducive to development, the theory can be explicitly adopted. When not, and the need for land reform becomes pressing, theory is forgotten and practical matters come to the fore.[14]

Nevertheless, the royalty system has economic effects and was not simply a practical impediment to the industry. The most obvious economic effect is the existence of the royalty payment itself, and we shall argue that it had an obstructive effect on the development of the industry quite apart from practical difficulties.

The royalty constituted a payment for access to the land by mining capital. In pursuit of surplus profits, capitalists could invest larger sums of finance, as it became available to them, even if all were reduced tendentially to a common level of profitability by competition. This would be encouraged by landowners whilst that capital flowed onto their own land and they would share in the surplus profits with the mineowners. In conditions of plentiful supply of small capitalists, the disincentive to invest created by the need to share surplus profitability with the landowner would be slight in aggregate because of competition between capitalists.[15] The situation is completely different once mining capital is simultaneously flowing onto a variety of lands. Then each landowner can potentially seek to share in the surplus profitability of any (extra) capital that flows onto any of the lands involved. Depending on the power of the landlord *vis-à-vis* the capitalist, something determined in part by competitive conditions and in part by social (e.g. legal) ones governing the conditions of access, there will be disincentive to expand capital extensively across many lands (or more exactly properties) including the case where this involves technological change such as mechanisation. Any surplus profits thereby generated are subject to appropriation by landlords directing mineowners away from large-scale extensive, mechanised mining. It might be thought that the mineowners with extra capital could purchase the necessary land. This merely serves to displace the problem to the price of land, the capitalised value of anticipated royalty (and surface) earnings, which would be forced up. This explains perhaps why few mineowners purchased coal land outright. Casual observation seems to suggest this may have changed slightly towards the end of the thirties, the interwar years being the period when the economic power of the landed interest began to yield under the burden of taxation. Nevertheless, by the interwar period, the system of landed property had become far from conducive to the reorganisation, mechanisation and concentration of the industry.[17]

The Miners' Attitudes to Royalties

In the latter half of the nineteenth century there was much agitation over the unequal ownership of land in Britain and the demand for it to be nationalised was raised.[18] A survey undertaken in the 1870s to show how widely dispersed landownership was proved quite the opposite and no other has been undertaken subsequently.[19] In this context, it is not surprising that the trade union movement should take up the demand for the nationalisation of the wealth underneath the land as well as its surface use. The Miners National Conference in 1899 welcomed the Royal Commission on Mining Royalties and expressed their hope 'that the outcome of its labour may be the full and, complete restoration of the mineral to the State'.[20] The first annual conference of the Miners Federation of Great Britain (MFGB) raised on its agenda 'Mining Royalty Rents: what position shall this Federation take with regard to the Commission in getting evidence etc.?' At the TUC of 1891 a motion was passed supporting the nationalisation of the minerals and mines calling for a 'Bill for restoring to the country its property in minerals and metals in terms of existing statute laws'.

Despite this apparent unity of purpose there were divergences of opinion amongst the miners' representatives. To place this in perspective, it is necessary to be aware that these differences went far beyond the question of royalties and centred upon much more significant matters. On an economic issue such as the eight-hour day, for example, the districts of the north east were opposed to legislation and this entailed their remaining outside the MFGB. Politically there was a strong tradition of liberalism amongst the miners which was also associated with a conciliatory attitude towards the relations between capital and labour expressed in the wish to work for the mutual benefit of both. Certainly at the time of the Royal Commission on Mining Royalties, there was no unity of opinion as is seen by the evidence presented by miners' representatives in answering questions from the Commissioners.

The diversity of opinion presented by the miners is indicated by the Commission's attempts to get Ben Pickard to attend as a witness. Pickard, President of the MFGB, had been out of town when first invited by letter and had declined to give evidence on the grounds that he could add nothing beyond what had been offered by Cowey, Haslam, Woods and Aspinwall who had appeared in the meantime. His unwillingness to appear may have reflected a wish to avoid adjudicating between his colleagues' opinions, so delicate were the ties holding the MFGB together. Alternatively, he may have thought it a waste of time.

It is possible to read the relevant Minutes of Evidence as a patronising exercise by the Commissioners, in which they indulged themselves in teaching the miners the elementary principles of economics and private property.

In terms of the effects of the royalty system, there was amongst the miners some, but very little, disagreement. The evidence of the miners as a whole is best considered in terms of a divergence of experience than of opinion, although there was also some uniformity of experience.[21] Stoppages because of excessive royalties or intransigent landlords were rare, but not unknown, and difficult to prove not to be due to other factors.[22] Many of the miners, however, commented upon a distributional conflict between royalties and wages, observing in particular that the mineowners often argued that they were unable to pay higher wages because of the level of royalties.[23] This was a source of industrial conflict between capital and labour over sums that were quite often lower than those paid for the royalties.

A more sophisticated relationship between wages and royalties was also revealed.[24] The payment of royalties meant that less advances were available to pay wages, even if labour could be profitably employed, because the royalties were a first call upon the finance of the capitalist. This was most disastrous for the miners when a mineowner was bankrupted so that even the wages for work already done might be lost.

In terms of foreign competition, the miners presented the case that royalties raised the price of coal and this led to the loss of export markets.[25] In answer to this, the Commissioners learnt to prepare a trap. Asking whether this view was based on the belief that royalties were higher in Britain than abroad, the miners, on answering in the affirmative, were implored to inform their colleagues that this was not so and that they should abandon their discontent with the private ownership of royalties. Some recognition was made of the practical difficulties arising out of the private royalty system. These concerned coal barriers, wayleaves and the need for collective pumping of water. Not surprisingly these could be more extensively documented by engineers and mineowners.

Where there was more disagreement amongst the miners was in what to do about the royalty system. Not all agreed that state ownership was desirable although this was the majority opinion.[26] Those in favour of state ownership were of the view that this would directly or indirectly benefit the miners' real wages, although Smillie also saw the state as using its powers as leaseholder to guarantee provisions for safety. The

Commissioners could, however, take the argument one step further. If the miners, or anybody else for that matter, were to benefit from state ownership of the royalties, it could only be at the distributional expense of the current royalty owners according to the compensation they were paid. This led to discussions over what compensation should be paid, with 'harder' or 'softer' lines being adopted by the miners, the content of contributions necessarily reflecting ethical judgements over who should be entitled to the produce of the land.[27] Keir Hardie, for example, merely wished to compensate widows and their children whilst Woods argued for full compensation to be paid.

To a great extent, these discussions were conducted on the basis of the royalties existing as a geologically determined differential rent and as a payment for the mineral independent of the form of property ownership. The same is true of discussion of proposals to abolish royalty payments, not necessarily through state ownership. In effect, abolition of royalties as a whole was seen as the generalisation of an individual royalty owner foregoing payment from the mineowner. This allowed abolition of royalties to be discussed as if it were simply the reduction of a cost without the necessity of proposing an alternative method of determining who would have access to the land for the purposes of mining. The effect of abolition could be discussed purely distributionally according to whether the lost royalty was rediscovered in the profits of the mineowner, the wages of the miner or in a reduced price of coal. This led the miners to accept that, although they might be reduced, the royalties could not be abolished without creating unequal rewards within the industry. Those that refused to recognise this were dismissed by the Commissioners, following a patient lesson in Ricardian political economy supplemented by the suggestion that the buying and selling of mineral rights might be best left to privately interested parties.

In discussing the abolition or state ownership of the royalties both the Commissioners and the miners tended to see the royalties either as a revenue for redistribution or as a cost to the system as a whole as if it were a cost to an individual mine. That the royalties were not only a differential revenue but also a differential cost was not entirely neglected. Young argued that the effect of abolishing royalties would be to disadvantage those mines paying the lowest royalties which would then be forced out of business. As he thought this would apply to Northumberland whose miners he represented, he opposed abolition of the royalties because of the effect on unemployment in his own district. It is not the self-interestedness of this argument as against other miners' interests that is so important as the rare focus that it contains upon production. Essentially it recognises that the royalty system has an effect on where mining takes place and even suggests that it sustains production at the less efficient sites. Paradoxically, this is so for the least progressive of the miners' representatives. Those demanding state ownership on distributional grounds were unable to recognise the effect of the royalty system on production.

This had not so much changed as become irrelevant by the interwar period. Out of adversity and diversity, the miners had created, by the end of the First World War, a united and militant union whose campaign over wages and hours took the political form of the demand for nationalisation. The arguments for nationalising the royalties as a means to reorganise the industry could be left to be formulated by the liberal bourgeoisie. The miners set themselves the task of mobilising support by relying more upon distributional arguments. The money paid to the royalty owner could more equitably be used to provide the miners with higher wages, shorter hours and better working conditions, particularly since the royalty owners often owed their position to the fortuitous acquisition of the coal centuries ago when its presence was not even suspected. Nevertheless, the primary focus of the miners was upon the mines rather than the coal, although the nationalisation of the first almost certainly entailed the nationalisation of the second.

In the wake of the General Strike of 1926 the miners were to be decisively defeated and effective schemes for reorganising the industry were set back for twenty years. During the thirties, the union was slowly and painfully rebuilt, with the threat of a national wage strike reappearing in 1935.[28] Nevertheless, the union was too weak to press for nationalisation and opposed schemes for reorganisation – of which royalty nationalisation might have been a part – for fear of the results on unemployment in the already depressed conditions of the industry. The royalties were finally nationalised in 1938 with little effective opposition other than from the royalty owners themselves. Sixty-six million pounds was paid in compensation. Even then, existing leases were to be honoured. It was only with the nationalisation of the mines after the Second World War that the immediate impediment to the development of the industry presented by access to the land was finally removed.

The role of the royalties was by no means over. The royalty owners had gone but compensation remained to be paid. Without any justification whatsoever the bill for compensation was treated as a capital liability of the newly formed National Coal Board (NCB) upon which interest had to be paid to the government. The royalties as an asset were finally written off in 1973, at a value of £52.4 million. As the original compensation had been £66.5 million, the NCB had paid to the government £14.1 million plus interest over seventeen years on the rest. In Britain, all of the nationalised industries have been controlled to some extent by financial mechanisms in which internal financing has played an important role. Whether these industries show a deficit or surplus on their accounts may give little or no indication of their performance. It may, for example, reflect the discovery of North Sea gas or rapid improvements in technology. Yet the accounts necessarily play an important role in the investment programme of the industries, restricting them to the extent that they reveal a deficit. For a variety of reasons the NCB has tended to show a deficit and its investment programme has been restricted.[29] This can only have been worsened by the need to pay for the royalties. Moreover government policy has also been to restrict price increases of coal, further limiting the potential for generating funds for investment.

The overall result is a remarkable historical irony. Towards the end of the nineteenth century, the miners began to press for the nationalisation of the mines and the royalties in the belief that the industry would be relieved of the burden of payment for the minerals. In fact, neither nationalisation was immediately achieved and, by the thirties, the industry was organised into a state-managed cartel in which the direct costs of the royalties and the indirect costs of the royalty system were supported by higher prices through output restriction.[30] After nationalisation of the industry and royalties, the costs of the latter were to a large extent borne by the NCB without any mechanism for a compensating increase in price. What the miners hoped to abolish was paradoxically created by nationalisation!

Conclusions

What conclusions are to be drawn from this analysis? The first is that the role of landed property, or rent theory as it is improperly termed, is not to be treated in distributional terms alone or even primarily. The effect of landed property goes deeper than the monopoly of landowners whose relative strength determines the size of their rental reward. It is necessary to examine historically and concretely the system of landed property and its situation relative to other economic and social relations. For the British coal industry, the system of landed property appeared conducive to the industry's development for a period prior to the First World War but rapidly became an impediment subsequently, despite no strengthening in landlord power nor in the level of royalties.

It follows that no general theory of rent is acceptable in which economic development or, more narrowly, distribution is determined by abstract principles. The effects of landed property can only be understood by considering the conditions of access (of capital) to the land and these are historically specific. It is neoclassical (and Ricardian) economics that examines rent as a condition of the land. It is an abstract Marxism that relates it to the conditions of capital accumulation (as a simple product of differing organic compositions of capital). As a distributional category of class conflict, rent is squeezed out of the bringing together of capital and land in the market place. A critique of these views, and arguments for the historical specificity of landed property, has been presented in the debate between Ball (1980) and Fine (1979 and 1980). The poverty of a general theory of rent has been revealed here by its inability to account for the place occupied by landed property in the changing fortunes of the British interwar coal industry. At a practical level, the problems were extensively recognised so that practical problems were seen to require practical, and not theoretical, solutions even to the extent of nationalisation of the royalties. When, as in the previous century, all is well with the world of landed property, theory reigns supreme and dictates that landed property has no effect so that there is no point in modifying it. As soon as problems begin to overwhelm, it is the theory that must be forgotten in order that practical solutions can be proposed.

Finally, let us consider the attitudes of the representatives of labour. Essentially, they adopted a distributional perspective which was ideologically progressive and politically mobilising by attacking the parasitical weight of landed property. This weight was not so great in distributional terms. Royalties formed somewhere between five and ten per cent of the price of coal and wage costs took up about eighty per cent. Royalties could have done something but not much to relieve wages and employment directly. On the other hand, they could have made a major contribution to the capital reconstruction of the industry quite apart from the need to remove obstructive effects that they had upon the investments that were made.

It would be easy, but wrong, to suggest that the limited ethical and distributional perspective that the miners adopted towards royalties was a factor contributing towards their defeat in 1926. There is no such simple relationship by which political practice can be read off from theoretical analysis. A penetrating understanding of the role played by royalties was far from necessary to guide the miners. What was required was clear enough and anti-royalty (and mine-) owner feeling could hardly have been more extensive. Paradoxically, where the miners' understanding let them down was in their very moment of success – when the industry was nationalised. It would have been possible at this juncture not only to have placed as first call upon the industry immediate economic conditions such as wages and hours but also to have struggled for a more progressive form of nationalisation itself. If nationalisation was seen as the solution to the problems of the interwar years, it certainly did not explain them. Without such an explanation, even the royalties were to continue to effect the progress of the industry in the ghostly form of compensation payments. This was itself symptomatic of wider mechanisms of control over the industry which could have been better understood and combatted if those that they replaced had themselves been fully comprehended.

Notes

1. Research into the British coal industry has been supported by the Social Science Research Council, the Leverhulme Trust, the Nuffield Foundation and the Central Research Fund of the University of London.

2. For a standard account of the industry leading up to the First World War see Taylor (1961 and 1968). For histories covering both periods see Kirby (1977) and Buxton (1978), for example.

3. The poor performance of the British industry is also revealed in transportation, distribution, haulage underground, mine layout, marketing, coal preparation etc.

4. Not all of these characteristics have always been considered failures. See Buxton (1970), for example, who argues that there was little to be gained from concentrating production on fewer mines. For a critique of this view on its own terms, see Evans and Fine (1980a).

5. The classic statement of collective, personal failure is due to Lord Birkenhead: 'it would be possible to say without exaggeration that the miners' leaders were the stupidest men in England if we had not had frequent occasion to meet the owners.' For a critique of the thesis of entrepreneurial failure over mechanisation in the interwar period see Evans and Fine (1980b) and for a critique of McCloskey's (1971) examination of the period prior to the First World War, see Fine (1982d).

6. The extent to which neo-classical economists wish to explain economic development in terms of given technical conditions, such as geology in the case of mining, borders on the fanatical. It is, however, generally but reluctantly admitted that, whilst Britain had inferior conditions to the United States, the same could not be said of its conditions relative to European competitors. It must also be recognised that the geological conditions known and worked are not exogenous but subject to both discovery and use. See Fine's (1982d) critique of McCloskey (1971).

7. For such an analysis, see Fine, O'Donnell and Prevezer (forthcoming).

8. Supporting evidence for the description of this section of the ownership and leasing of coal land is to be found predominantly in the Reports and Minutes of Evidence of the three Royal Commissions (1893, 1919 and 1926) that considered the question, the first one exclusively so. Further analysis of the effect of the royalties other than in references already made is to be found in Fine (1978), Fine (1982b) and Brunskill, Fine and Prevezer (1982).

9. The great landed aristocrats of Britain are of course a cliché of its history. See Massey and Catalano (1979).

10. See Fine (1982d).

11. A problem not mentioned by the Scott Committee but of great importance and of interest because of its relation to the control of labour is as follows. It is important to recognise that coal is not a homogeneous product as it yields a vast variety of different characteristics, including size. At this time and into the period of nationalisation, there was a premium on large coal. In relation to the royalty owners, the mineowners had an incentive to extract large coal and leave the small to the extent that royalties were paid by the ton or some form of equivalent. By the same token, to the extent that the miners were on tonnage piece rates, they would fill tubs with small coal where allowed and, if fined for doing so, would neglect small coal. As piece rates were prevalent with fines in one form or another, the result was the wastage of small coal.

12. When the royalties were nationalised in 1938, it became clear that the number of leases per mine at that time would need to average between ten and fifteen even on the basis of the existing fragmented and small scale layout of the British mines.

13. As Marx observes: 'The radical bourgeoisie . . . therefore goes forward theoretically to a refutation of the private ownership of land, which, in the form of state property, he would like to turn into the common property of the bourgeois class, of capital. But in practice he lacks the courage, since an attack on one form of property – a form of the private ownership of a condition of labour – might cast considerable doubts on the other form. Besides, the bourgeois has himself become an owner of land' (Marx 1969: 42–3). This last condition is not substantially met in the British coal industry. For an exposition of Marx's analysis of the state ownership of land, see Fine (1982a).

14. For a consideration of the neglect of the significance of the system of landed property in the history of economic thought, see Fine (1982a and 1982c).

15. This characterises the conditions in Britain, certainly from the beginning of the railway age until the accelerating erosion of these conditions around the turn of the century.

16. See Massey and Catalano (1978).

17. The argument here has followed, for the historical circumstances governing landed property and mining in Britain, the theory of agricultural rent associated with Marx. See *Capital III* Part IV and *Theories of Surplus Value II*. Recently, interest in Marx's theory has been revived in the context of urban crisis. Marx's analysis has often been treated as if it simply concerned distribution of surplus value between landlords and capitalists (and workers) and has otherwise been rejected where it is recognised that it does not so conform. Alternatively, Marx's theory has been interpreted as containing an abstract and ahistorical validity independent of the conditions governing access to land in a paradoxical reconstruction of the property of neoclassical (and other) theory in which rent is independent of such historical relations. For one interpretation of Marx's theory see Fine (1979), and for a rejection of the two other positions and all general theories of rent, see the debate between Ball (1980) and Fine (1980).

18. See Eldon Barry (1965).

19. This was called the New Domesday Survey. See Bateman (1883) for a presentation of much of its results. If Scotland is anything to judge by, only a little had changed in the passage of one hundred years. See McEwen (1977) and Clark (1981). Royalty ownership was more concentrated than land, see Brunskill *et al.* (1982).

20. See Arnot (1949) for the narrative contained in the rest of this paragraph.

21. Most of the miners' Evidence is to be found in the second volume of Minutes of Evidence.

22. See Appendix IV to the Final Report.

23. See the evidence of Wilson, Small, Cowey, Aspinall and Woods. Keir Hardie believed that royalties only had an indirect effect on wages through higher employment brought about by lost export markets.

24. See the evidence of Small.

25. See the questioning of Aspinwall.

26. Those against state ownership included Aspinwall, Brown, Cowey and Haslam.

27. An exception is to be found in Smillies' argument that some of the costs saved by a more efficient pumping could find their way into the pockets of the miners.

28. See Arnot (1961).

29. In this it contrasts with the state-owned Central Electricity Generating Board which has been able to pursue a massive and unnecessary investment programme, including provision for nuclear power. For a discussion of control of the nationalised industries in the UK, see Fine and O'Donnell (1981) and Fine forthcoming for a discussion of nuclear power.

30. See Kirby (1977) and Fine, O'Donnell and Prevezer (forthcoming).

References

Arnot, R.P. (1949) *The Miners 1881-1910: A History of the Miners' Federation of Great Britain*, Allen and Unwin, London

—— (1961) *The Miners in Crisis and War, from 1930 Onwards*, Allen and Unwin, London

Ball, M. (1980) 'On Marx's Theory of Agricultural Rent: A Reply to Ben Fine', *Economy and Society, 9*, 3, 304-26

Bateman, J. (1883) *The Great Landowners of Great Britain and Ireland*, 4th edn., Harrison, London, reprinted 1971 by Leicester University Press with commentary by D. Spring

Brunskill, I., B. Fine and M. Prevezer (1982), 'The Ownership of Coal Royalties in Scotland', mimeograph

Buxton, N.K. (1970) 'Entrepreneurial Efficiency in the British Coal Industry Between the Wars', *Economic History Review, 23*, 3, 476-97

—— (1978) *The Economic Development of the British Coal Industry*, Batsford Academic, London

Clark, G. (1981) 'Some Secular Changes in Landownership in Scotland', *Scottish Geographical Magazine, 97*, 1

Coal Association (1920) *The Evidence on Nationalisation of Minerals and Royalties Critically Examined*, pamphlet

Eldon Barry, E. (1965) *Nationalisation in British Politics*, Jonathan Cape, London

Evans, T. and B. Fine (1980a) 'Economies of Scale in the British Interwar Coal Industry', *Birkbeck Discussion Paper*, no. 75

—— (1980b) 'The Diffusion of Mechanical Cutting in the British Interwar Coal Industry', *Birkbeck Discussion Paper*, no. 76

Fine, B. (1978) 'Royalties and the Interwar Coal Industry', *Birkbeck Discussion Paper*, no. 62
—— (1979) 'On Marx's Theory of Agricultural Rent', *Economy and Society, 8*, 3, 241–79
—— (1980) 'On Marx's Theory of Agricultural Rent: A Rejoinder', *Economy and Society, 9*, 3, 325–31
—— (1982a) *Theories of the Capitalist Economy*, Edward Arnold, London
—— (1982b) 'Landed Property and the Distinction Between Royalty and Rent', *Land Economics, 58*, 3, 338–50
—— (1982c) 'The Historical Theory of Rent and Price Reconsidered', *Australian Economic Papers*, June
—— (1982d) 'Landed Property and the British Coal Industry Prior to World War I', *Birkbeck Discussion Paper*, no. 120
—— (forthcoming) 'Nuclear Power' in B. Fine and L. Harris, *The Peculiarities of the British Economy*, Lawrence and Wishart, London
—— and K. O'Donnell (1981) 'The Nationalised Industries' in D. Currie and R. Smith (eds.), *Socialist Economic Review*, Merlin, London and, as revised, in B. Fine and L. Harris, *The Peculiarities of the British Economy*, Lawrence and Wishart, London (forthcoming)
—— and M. Prevezer (forthcoming) 'Coal Before Nationalisation' in B. Fine and L. Harris, *The Peculiarities of the British Economy*, Lawrence and Wishart, London
Jevons, W.S. (1865) *The Coal Question: An Inquiry Concerning the Progress of the Nation and the Probable Exhaustion of our Coal Mines*, 3rd edn. (1906), Macmillan, London
Kirby, M.W. (1977) *The British Coal Industry 1870–1946: A Political and Economic History*, Macmillan, London
McCloskey, D. (1971) 'International Differences in Productivity: Coal and Steel in America and Britain before World War I' in D. McCloskey (ed.), *Essays on a Mature Economy, Britain after 1840*, Methuen, London
McEwen, J. (1977) *Who Owns Scotland?*, E.U.S.P.B., Edinburgh
Marx, K. (1969) *Theories of Surplus Value, II*, Lawrence and Wishart, London
—— (1972) *Capital*, Lawrence and Wishart, London
Massey, D. and A. Catalano (1978) *Capital and Land: Landownership by Capital in Great Britain*, Arnold, London
Mining Royalties (1893) *Report of the Royal Commission on Mining Royalties. Final Report with Minutes of Evidence and Appendices*, C. 6980
Nef, J.U. (1932) *The Rise of the British Coal Industry*, 2 vols., Routledge, London
Reid Report (1945) *Report of the Technical Advisory Committee on Coal Mining*, Cmnd 6610
Samuel Report (1926) *Report of the Royal Commission on the Coal Industry*, Cmnd 2600
Sankey Report (1919) *Reports of the Royal Commission on the Coal Industry with Minutes of Evidence and Appendices*, Cmnd 359–61
Scott Report (1919) *Ministry of Reconstruction. Third Report on the Acquisition and Valuation of Land for Public Purposes of Rights and Powers in Connection with Mines and Minerals*, Cmnd 361
Taylor, A.J. (1961) 'Labour Productivity and Technological Innovation in the British Coal Industry, 1850–1914', *Economic History Review, 14*, 1, 48–70
—— (1968) 'The Coal Industry' in D.H. Aldcroft (ed.), *The Development of British Industry and Foreign Competition, 1875–1914: Studies in Industrial Enterprise*, Allen and Unwin, London

PART THREE:
THE DEBATE OVER MARX'S THEORY OF RENT

7 A MARXIST APPROACH TO URBAN GROUND RENT: THE CASE OF FRANCE

Alain Lipietz

The purpose of this paper is to present the outcome of research which started in the early 70s in the context of the French Housing Ministry.[1] At that time and for a 20 year period, France had experienced unprecedented industrial and urban growth and the urban question became a major issue. The tools of main-stream economics were proving themselves ineffective and even deceitful. Hence a tolerance developed for non-orthodox and even Marxist theories.[2]

Nowadays the economic crisis has moved the urban land question to the background, as the oil rent problem has become a burning issue. On the other hand, the phenomena related to urban landed property carry the trade-mark of the specific social formation in which they occur. Thus, the present study cannot be transferred to other industrialised countries without great caution. Our purpose is simply to show to what degree the Marxist theory of agricultural ground rent can be useful for our understanding of the laws of urban land prices, provided that it is not naïvely transferred word for word.[3]

The first section is devoted to a critique of the main-stream approach, the second outlines the theory of what we will call 'urban land tribute'; the third section studies this tribute's sources and the fourth analyses its variations in space and time.

Methodological Questions

Empiricism

At the time, a first approach was in favour of watching, comparing, measuring. It seemed legitimate. Computers made the task easier. They could 'digest' large quantities of data related to the sample's features. And they produced the following law: 'The price of land is equal to 1/8 the product of the cost-price per square metre of surface and of the urban occupation coefficient' (Dutailly 1970).

The scientific sterility of this method, if we take it as the only 'positive' one, is evident: it does not teach us anything about an explanation, it only reflects the effects of contigent conditions imposing themselves through permanent mechanisms over which the theory has nothing to say. Moreover, we know nothing about the conditions of validity of such a law, neither do we *a fortiori* learn anything about ways of modifying it. This law gives us a picture, 'a quiet reflection of the world of *phenomena*' (Hegel). However our interest is in the essence, and we have to explain the phenomenon by means of concepts coherent with those which have permitted us to illuminate other aspects of reality. Therefore we have to start from a theoretical framework and test its relevance for the particular field under study.

Marginalism

Most theories of urban land prices are influenced by the 'dominant theory': marginalism. They approach urban land as a 'good in itself', desirable to the consumer because of its usefulness.

Let us not study the mathematical problems raised by the introduction into the general equilibrium theory of this good which, besides quantity and price, possesses another quality: location. We will look mainly into the epistemiological basis of this approach.

For these theories, land value is based on the utility of locating one's housing on such and such a plot (in fact at this or that distance from the centre). Land being a scarce good, it is thus subject to the theory of 'optimal allocation of scarce resources'.

The optimum to be achieved is a distributional one. Since the stock is fixed, the optimum depends only on the 'utility function' in the user's mind. For Wingo (1964) and Mayer (1965) the trade-off is limited to a reduction in the generalised transportation cost of the home-job trajectory: a minor aspect but which has the credit of being measurable. Alonso (1964), in a more cautious way, makes utility dependent on the distance and the plot's area, without specifying the shape of the function which can have extremely varied outcomes, depending on the 'elasticities of substitution'. Maarek's model (1964) gives to empirical data the task of specifying the function's parameters.

The theories have all the shortcomings of the marginalist school, over-emphasised for the case of 'distributional optima'. They are theoretically based on psychology, regarded as the irreducible root. To develop the numerical model we need, in addition, to suppose a particularly productivist psychology (Wingo) or else either stay in uncertainty (Alonso), or fall back into empiricism (Maarek).

The most bothering feature of this approach is that it implicitly or explicitly (Mayer 1965) involves a separation between the price of the building and that of its plot, as if one could freely combine them *à la carte*.

However, 'urban land' is never demanded for itself but for the access that it allows, when it is at one's disposal, in the sphere of economic and social relations. In a social formation, this possibility of disposing of land expresses itself in some legal manner (ownership, leasing . . .) which can give rise to a transaction but a transaction which relates (and this is the essential point) not only to abstract land, but to a particular access granted, within the economic and social sphere, by the legal claim over a given plot.

Now, in capitalist social formations, to speak of a particular situation implies mainly (but not only) speaking of a particular situation with respect to the general movement of capital.

We are thus forced to leave the heaven of 'utility functions' embedded in the citizens' minds, and to anchor ourselves on the concrete ground which is subject to competition between capitals in their attempt to valorise themselves. More precisely we will study urban land prices in so far as they relate to the capitalist production of housing (CPH).

The problem, then, presents itself in the following way: to understand how the access to a legal claim on land is priced as a condition for the valorisation of capital. Now, this legal claim is in the possession of landowners who can therefore levy a 'tribute' on the circulation of capital. For the case of agriculture, the laws which govern this tribute were analysed by Marx,[4] basing himself on (but also correcting) Ricardo. Provided we do not attempt simply to transfer this analysis (which was left by the author as a very rough draft), we can use it for the study of the urban ground problem.

Towards a Theory of Urban Land Tribute

In order to study land prices as a tribute levied by landed property on the process of capitalist production and circulation, it is imperative to address two preliminary questions. What is produced? Which capital is set in motion? This seems evident. Yet numerous Marxists miss this point and ask a different and wrong question: 'Where is the rent?' As far as we are concerned, we will stick to the principle: start from production.

Suddenly, things do not appear as simple any more. What is in fact produced when housing is built? A shelter? Not only that.

The Built Environment as a Product

While wheat in Marx's and Ricardo's fields was a simple product satisfying a precise social need with minor qualitative modulations and merely susceptible to variations in yield per acre or in its transportation cost, the dialectic between product and the need and the multiple articulation of the 'instances' must be taken into account in the consideration of the 'products' that spring up on urban ground.

First of all, land is 'urban' only in so far as it is the base of urban society, which is the dominant lifestyle of capitalism.

This agglomeration is a rigid and constraining structure; it is not a functional organisation but the manifestation in space of the mode of production in the context of a historically-determined social formation. Just as capitalist 'civil society' is an antagonistic confrontation of social classes, so too the urban agglomeration is a deployment of the economic and social division of space (ESDS).

The Economic and Social Division of Space. The economic and social division of space is the spatial effect on the agents' practice and on the structure of their lifestyle of the totality of the social formation's structure (with its past). It reflects the effect of all instances (economic, political, ideological) and therefore it will not let itself be reduced to some economistic and mechanistic model. Nevertheless, the reality of its own presence in urban practices powerfully contributes to the maintenance of the social formation while exacerbating its contradictions. Particularly, 'the value of the home-job trip' will not be the variable which controls the modulation of land price, but instead the location within the ESDS, which is an infinitely more complex determination.

The economic division of labour which prevails at the level of productive forces and relations of production reappears in the economic division of space in the distribution of factories, offices and housing. The social division of labour which lies at the level of social relations of production reappears in the social division of space: here live the employers, there the engineers, there the workers.

The constitution of the SDS is a very complex phenomenon. It is first made possible, and determined in its spatial form, by the state of the urban transport system. It is then confirmed by the differentiated treatment given by urban administrations and public services. This is an immediate consequence of the dominant class's preferences. But it is also a domination of urban forces over economic ones: rich neighbourhoods welcome only rich buildings, 'public consumption of space' can only be secured on less expensive land. Moreover, the SDS is socially and consciously wanted; the most typical case is that of racial segegration: the arrival of black workers making middle-class whites move out. The SDS is also the object of a specific political practice. Urban planning becomes a strategic element in the maintenance of order: this 'police urbanism' being clear in the Paris of Hausmann. The SDS is finally re-affirmed legally and administratively by 'zoning' decisions.

All these indications do not pretend to constitute the theoretical notion of the ESDS. They simply attempt to throw enough stones at the pond that reflects the peaceful image of a radiocentric city inhabited by homogeneous wage-earning citizens who sensibly weigh the substitutability of their precious time and money. Behind the 'map of land prices' we have to read the geological map of the economic and social uses of urban land.

Inside and around the city we thus have a series of users, ranked in decreasing order of ability to pay for location: the superior tertiary sector (banks, high level trade), housing (ranked by the users' incomes), industry and finally agriculture. Again this classification articulates with the reality of geographical disposition historically given by numerous determinations.

And what about the public use of urban space? This is where we have to approach the concrete production conditions of the built environment: one way is to consider a private capitalist who uses land as a way to produce a good (housing) or as a way to valorise his capital (a factory or a store), another is when the political power, whose function is to assure social cohesion, intervenes by creating the general conditions of social reproduction, constructing roads or the necessary public buildings. In any area where it operates, the urban land tribute that will have to be paid will not be determined in an endogenous manner by its own activity but only by the prevailing use in that particular area.

Therefore we will hold that public use of land does not create a specific form of urban land tribute other than the one provoked by the private use that it induces. If we temporarily put aside the dominant (but less important) special case of the superior tertiary use, we are brought back to the problem of land use for housing production which is the principal element of the built environment.

The Housing Product: A Special Commodity. We can analyse a commodity in two different ways: as a need satisfied or as a product sold. The line of reasoning which starts from needs is dangerous and may be even mystifying in a social formation ruled by the capitalist mode of production, where the aim of production is the growth of capital. This growth is, of course, subject to the realisation condition: in order for something to have exchange value, it must also have use value. But it is capital which, from the set of possible needs and desires, chooses those which it is profitable to satisfy.

Nevertheless, housing cannot be thought of as just another commodity, for two reasons:

(1) Housing, even when it is reduced to a simple den, is an essential element in the reproduction of the labour force; it therefore affects its value as a commodity.

(2) Private housing is the 'structuring focus' of consumption in capitalist urban society. It is around and through housing that 'household consumption' is organised and modulated so that it regulates the growth rate of expanded reproduction.

It is through a typical type of housing more than through the myth of equal access to elementary school that the belief in social homogeneity in 'Civil Society' imposes itself on antagonistic classes. This aspect of the role of housing has become crucial since, after World War II and in all industrial countries, the 'intensive accumulation scheme' has become generalised. It was based on very fast productivity gains and on mass consumption of automobiles, household appliances etc.[5] This is why the den, which was characteristic of primitive accumulation, has been progressively replaced, since the turn of the century (in France mostly since the Second World War), by a homogeneous model of housing, only slightly variable in its technical characteristics (layout, services), therefore barely variable in its construction cost per unit of area. Thus, it is a good whose production cost is relatively invariable which is offered to a set of social strata whose purchasing power varies from 1 to 10.

Now we have to bring together these two results:

(1) The social and economic division of space is a given that cannot be attributed only to economic considerations. It can be considered as a given for urban development, though development nevertheless leads to its modification. As a habitat, it is geographically much differentiated by income level.

(2) Housing production provides a homogeneous product for a demand which is itself highly differentiated by income groups.

Now the good sold is not the building but the housing, that is a building localised very specifically in the ESDS.

It is in the lag between what is produced and what is sold that we shall find one of the several origins of urban land tribute. But first we must deal with the question: who (which capital) produces housing, and how?

Capitalist Housing Production

Who produces built environments and particularly housing? This question has produced much confusion and is the origin of numerous misunderstandings. We have, therefore, to underline vigorously this obvious fact: the built environment is the product of the activity of construction firms – just as cars are produced by automobile firms and a nation's armaments by arms manufacturing firms. We have to add that all construction firms in France are private. Therefore, at least formally, the built environment in France is entirely produced under capitalist conditions.

Nevertheless, there is an important difference between this and other products of capital: in France generally, money-capital which is intended to turn through the production process into a house or road is not initially in the hands of capitalist contractors (the *maîtres d'oeuvres*). It is not their property. This capital is normally in the hands of *maîtres d'ouvrages*, or 'developers' (*promoteurs*).

It is here that we have to bring in the basic distinction developed by Bettelheim (1975) between 'property' and 'possession'. The capitalist production process is the unity of two processes: a labour process (setting in motion productive forces: equipment and materials, human labour force, in pursuit of the actual production of use-values) and a valorisation process (an engagement of capital in pursuit of its own growth by surplus-value extraction). We speak of possession to designate the relation to the production unit of the agent who dominates the labour process, who 'sets in motion the productive force', and of property to designate the agent who dominates the valorisation process, that is to say, the one who has the power to direct labour to a particular activity and to dispose of its product.

In the capitalist mode of production, where the valorisation process dominates and informs the labour process, the agent who dominates the valorisation process also dominates the production process; he is therefore the real owner, even if he is not (e.g. in the case of a promoter who uses a loan) the legal owner.

In the present conditions of Capitalist Housing Production in France, there is a separation between the 'property' of capital which circulates productively in the CHP, which we will call 'development capital', and this capital's 'possession' in the production process, which is in the hands of construction firms, building entrepreneurs.

There are two reasons for this: (i) the final purchaser's lack of funds: it is thus necessary that a pre-financer advances the money capital to the entrepreneur and then transfers the housing-good to a capitalist (maybe himself) who will act as seller to the user; and (ii) especially the landed property problem.

In France, landed property has been, since the Revolution of 1789, scattered through all social groups. One of the conditions for housing production thus eludes the scope of any firm's production. No capital can produce it, because a 'building site' is not a product, but the inscription in space of numerous social relations. On the other hand, as many sociological studies have shown, the landowners, those who legally dispose of land, are generally not quick to get rid of it, no matter what economic advantages they would obtain. To get a free plot is usually beyond the scope of the building entrepreneur.

Thus, to produce housing is not only to engage capital in a building activity, it is to 'set up a development programme' where capital circulates under the *maîtres d'ouvrage*'s 'property' through the *maître d'oeuvre*'s 'possession'.

It is the role of the promoter to advance the money to 'buy' the right to build from the land owner, to advance the circulating capital put at the contractor's disposal, to pay the fixed capital (which is anyway the contractor's property) and to manage the stock of housing until its final sale. Such is the rotation cycle of promotional capital. The building firm appears then as a 'sub-contractor' of a formal firm: the development firm. The necessity to liberate a piece of land each time to re-engage the reproduction cycle of construction capital and to find a new partner each time, brings the building firm to limit the use of fixed capital which it is not sure of being able to pay for regularly. The firm will also try to maximise the role of circulating capital (specially wages) since it will be advanced by the promoter.

For these reasons the organic composition of capital is weaker in the construction industry than in other industrial branches. Now, the Marxist theory of value shows that, in the case of such a branch, a relatively higher surplus value is produced (relative to capital employed) than in other branches.

As for 'promotion capitals', they constitute a rather complex system (Topalov 1970) that will not be studied here because this system varies too much from one country to another and from one time to another (see Chapters 2 and 3). It compounds private capitals of very different origins and magnitudes, obeying various logics, and public aid and loans aimed at coping with the uneven access of households to housing. But we know enough to enter into the heart of the matter.

The Rise of the Urban Land Tribute

Let us imagine a capitalist aiming to collect a profit by engaging his capital in housing production. To simplify we will suppose he is a 'builder', that is both promoter and contractor at the same time, and that he intends to appropriate an 'average profit' determined by industrial activity as a whole. He knows that in a certain place within the social division of space he will be able to sell at a particular price. An obstacle to the operation: he doesn't own the land so he has to enter into a legal relation with the owner.

The differences from Marx's farmer are evident when the question is posed in these terms. While wheat production process is repeated from year to year with a timeless regularity, housing production takes place for a few months and it won't be restarted on the same site for many years. So the land tribute does not take the form of a regular rent, as in the case of agriculture, but it is a rather discrete transaction: the sale of the building site. Therefore the land price is not the simple capitalisation of a rent. It is the land tribute itself; it is the form which reveals the social relations between the owner and the promoter (the barter of the right of land disposal against a part of the profit), disguising it at the same time as the purchase of an economic good.[6]

While it is easy to measure the quantitative differences between harvests on different pieces of land, the difference between two housing products in two different points of the SDS is qualitative. Hence, the differences in 'productivity' between building sites don't have purely physical or economic origins but depend also on the sale price of housing at different points in the SDS (which is, let us recall, an effect of the totality of the social formation).

The nature of land prices appears clearer now. When the promoter 'buys' a building site, he doesn't advance his capital in the same manner as when he buys materials, machines or labour power. What the promoter is buying is a legal right which he doesn't pay for as a part of his productive capital, but as an advance on the surplus profit he expects to make (beyond the average profit that he reserves for himself). This is why land prices don't exist by themselves, but are created by the promoter's activities which give land a determined use; so capital may be multiplied on the same spot in a few years.

Let us summarise. The urban land tribute is the part of surplus profit which, for various reasons, promotional capital captures in capitalist housing production and which the landowner is able to appropriate. We say that the landowner 'is able to' because the promoter's ability consists in leaving the owner ignorant of the potentialities of his piece of land!

It remains to be seen why there is, globally, a surplus profit in capitalist housing production and what modulates this surplus profit and the land tribute from site to site and from year to year.

Where Does the Land Tribute Come From?

Why is there a surplus profit and how is it modulated? Economists generally don't put the question in these terms. In fact we will see that the question is rather formal. If we proceed in this way, it is to distinguish ourselves from prevalent concepts that appear to us as false and to introduce two different groups of concepts.

The Land Tribute's Structure According to Marx, Engels and the Marginalists

Marx's and Engels's Position. Marx distinguished two types of rent:

(1) That which is based on the relation between each plot and the worst cultivated plot: it is the surplus profit made by the farms which utilise the more fertile and the better-located plots (differential rents of type I). It is also the difference of surplus profits per acre due to the unequal distribution of capital over the plots (differential rents of type II). These differential rents (DR) are not due to the existence of landed property, but the latter is able to appropriate them.

(2) Those which are levied even on the worst plot, as a pure tribute of landed property on capital. It is the absolute rent (AR). Total rent (TR) paid on any plot is then of the form:

$$TR = AR + \Sigma DR$$

Then Marx asks the question:

Does absolute rent enter in the price as a tax received not by the State but by the land owner, that is to say, as an element independent of the value? (*Capital, VIII*, 142)

Of course, rent is fixed by economic conditions (competition, effective demand). How about its origin? Who pays the rent? Where is the value produced, the part of surplus value which is paid to the landowner? Marx begins by pointing out that rent can be compatible with the theory of value, i.e. that wheat can in fact be sold at its value, paying profit and rent at the same time. It is possible if the surplus value created by waged workers in agriculture exceeds the quantity sufficient to grant the average profit to capitalist farmers. We know that when it is so in other branches, the profit rate is levelled by an influx of new capitals which compete between themselves. Then relative prices are no longer regulated directly by value, but a system which is a 'transformation' of the system of values, the one of prices of production.[7] Here, landed property prevents the functioning of this mechanism, and the surplus over the average profit is used to pay the rent. We thus have:

Value > production price Existence of landed property	→	There is a surplus profit that is transformed into rent

This reasoning is a logical order, which defines the limits of ground rent in the framework of the labour theory of value. It is neither the real nor the genealogical order of the appearance of ground rent.

Marx immediately makes a reversal of the formulation:

> Due to the blockage caused by landed property, the market price should rise sufficiently to allow the plot to pay a surplus over the production price, that is to say, a rent. But, according to this hypothesis, as the value of goods produced by agricultural capital is superior to their production price, this rent, except for a case of which we'll talk later, is composed of the total or partial surplus of the value over production price. (*Capital, III*, 146)

In short, this is just fine! In the agricultural sphere, where it is necessary to pay a rent, value is above the price of production. As a matter of fact, the presence of landed property has clearly played the part of a restraint on capital accumulation, therefore imposing 'technologically' a surplus of labour value over the price of production in agriculture.

But it is not necessarily so. (We can think, for example, of rent in oil countries.) Marx knows it well and contemplates another source of land rent: the surplus profit obtained from the sale of a good whose price is fixed as a 'monopoly price', determined neither by the price of production nor by the value, but by the demand and ability-to-pay of the buyer. What this means is that the monopoly on land reflects itself in the transitory scarcity of a commodity (wheat, housing) which, by raising the cost of living, forces industrialists to subsidise the nobility through the workers' wages. It is the battle between Whigs and Tories concerning the monopoly on wheat.

How about housing? Marx dedicates only one sentence to the subject by referring to the 'absolute predominance of monopoly rent'. As for Engels (1969), he is emphatic: the rent which housing owners appropriate (in addition to amortisation and construction profits) is 'a cheat', and 'as soon as a certain average amount of cheating becomes a rule in any place, it inevitably has to find (in the long run) a wage increase as a compensation'. He thus refers to Marx's monopoly rent and not to the difference between price of production and value.

A Critique of Marx's and Engels's Positions. Here we would like first to challenge the 'AR + Σ DR' structure. For it is not enough to explain the sources of absolute rent! The same problem concerns differential rent. It is known that Marx's theory holds that the market price oscillates around the price of production (cost price + average profit) obtained in average social conditions of production. Now, this theory of differential tribute assumes that the base of the market price is the price of production imposed by the worst production conditions for wheat. So, the sum of profits and differential rents in total production of wheat exceeds the product: 'total capital employed × average profit rate'. In any case, and even without absolute rent, it is necessary to foresee a total surplus profit (in relation to the price of production), obtained in the sale of wheat, to pay for the differential land rents!

We are in the first place induced to ask a question: what is the source of the surplus profit which pays urban rent (and more generally the tribute)? That is, where does the excess value brought into play by the realisation of land-related production come from? And then: what makes surplus profit per acre different from one place to another?

Thus, the structure of price is not:

$$TR = AR + \Sigma DR$$

but: $$TR = AR \pm \Sigma DR,$$

the absolute rent being not the rent (or the tribute) on the worst spot, but on the medium one, so that the regulating price is determined by adding cost + average profit + average tribute, the 'differential tribute' being only differentiation of tribute.

Of course, the problem of the origin of the average tribute remains, just like the problem of the absolute rent in Marx. Here is the second criticism we would address to Marx. If the tribute is to be paid with surplus value not equalised through competition of capitals (because of the existence of landed property), this surplus-value could have been produced either within the branch (if value $\geq$ cost + mean profit) or outside (if value $<$ cost + mean profit). Thus the distinction between 'absolute rent' in Marx's sense and 'monopoly rent' is rather irrelevant[8] as we can see when computing the regulating price of the algorithm of transformation with a unit rate of profit and of the tribute per acre (Lipietz 1979b). Anyway, we shall have:

sum of profit + sum of rents = sum of surplus value

but the surplus value has no smell nor taste, and one can't say: this tribute in building production comes from the very production of buildings. Yet, there is the problem of producing this part of surplus-value and devoting it to landed property; that is what we call the source of tribute.

The task now is to create two groups of concepts:

(1) The sources of tribute, that is, the set of social relations which exist between the capital invested in building, agriculture or any other activity related to ground's disposition, and landed property, these social relations being expressed by different forms (ground price, rent) of *land tribute*;

(2) The modulation of the tribute by the articulation, with these relations, of other relations or practices of promoters, users, of the State, etc., articulations which cause various types of differentiation which will be designated by the term of *differential tribute*.

The Marginalist Position. It is interesting to note that the marginalist school explicitly acknowledges a structure of land price similar to that of Marx. We know the city's radius R (determined by its density and its population). We derive from its inhabitants' 'utility functions' a differential equation which expresses for all points the marginal increment that they are willing to pay in order to come closer to the city centre. It is a kind of situation-differential tribute at a distance r from the city centre. We have then to solve a Dirichlet differential problem – that is, it is necessary to know the ground price at the city's periphery, P_R. Therefore we have a structure similar to that of Marx:

$$P_r = P_R + \int_R^r \frac{dp}{dx} \, dx$$

We can compare this metric structure to that of a volcanic island (Mayer 1965): above a certain level, given exogenously at the periphery, the island's altitude increases regularly towards the centre according to laws concerning the equilibrium of slopes.

As far as the 'absolute tribute' P_R is concerned, its estimation varies according to different authors:

(1) $P_R = o$, an evidently unrealistic position of Maarek [1964] which builds the city in a sandy desert.

(2) $P_R = a$ (price of agricultural ground). This is Alonso's position; it is also that of Adam Smith which Marx takes up without much examination in *Capital*: agriculture being the predominant use of land, the housing absolute rent is the total tribute paid at the periphery of the city.

(3) $P = \begin{cases} a + b + c + d & \text{within the city} \\ a + c & \text{at the outside} \end{cases}$

This is Mayer's proposition: b is the development cost; c an 'anticipation rent'; d is a 'scarcity rent' on buildable plots. The 'volcanic island' will no more rise smoothly from sea level, but will rise at once, with a discontinuity, the 'Mayer Threshold' c + d; and it will produce waves (though we do not know how they damp out to reach a, the agricultural price).

It is useless to criticise this position theoretically since we don't agree even with Marx's position with regard to the implied structure of land prices, given our former criticism of the 'psychological' conception of urban price modulation.

When tested out, this conception clashes with some noted contradictory examples. The econometric adjustment made for Nantes has given unstable coefficients with 'distance', (which proves that 'the value of time' is not the same for everybody and that it is necessary to take into consideration at least the social stratification), with 'orientation' (which shows that the social typology of central districts of the city tends to spread towards the outskirts) and with time (which shows that ground prices depend on the general economic conjuncture). Above all, Vieille's study (1970) on Teheran shows that land prices are higher on uncultivated stony fields located north of residential districts, than on more fertile lands located south, where the poor districts are.

We are thus brought to think that land prices are determined from the centre towards the outskirts, the possible urban use entering in competition with the real agricultural use and being able to bring about a greater land tribute by itself. Therefore we have to think not of the 'volcanic model' but rather of the 'alpine model', with the rise of a granite core pushing waves all around.

The Structuring of Land Tribute by the Promotional System

In order to specify this idea, we have to deal with relations and practices which intervene in a concrete domain, that is, through and in the framework of a pre-given system, functioning as a whole, which specifies the elements and their relations. This domain is that of the functioning of the system of promotional capitals, more particularly in its confrontation with urban property but under the constraints imposed on it by the state of the wider circulation system of social capital, and that of effective demand.

The heart of the theory of urban land tribute is that each use of the land brings about its specific tribute in each specific situation. The problem then is: how is the land tribute fixed for a use? And how are the uses of land distributed?

In the first place, let us remember that the hierarchy of the economic and social division of space is dominated by the superior tertiary sector, which brings about a tribute corresponding to particular relations, and which is superior to that brought about by housing. On the other hand, agricultural use creates its own tribute which is generally inferior to that of housing.

The land required for housing is determined by the mass of capital invested in building, these capitals being divided into sub-markets by the level of public subsidy and thus requiring different rates of profit. This mass and these rates are determined by the conjunctural situation of the circulation of social capital: general rates of profit in industrial sectors, rate of interest, volume of mobile capital, monetary policy, and so on. The real estate sub-markets are defined by the confrontation of the promotional system and the structure of effective demand (the incomes of various social classes). They compete for the occupation of the pre-existing ESDS. The hierarchy of these sub-markets fixes the order of priority for the occupation of spots, from prestige tertiary to social housing. The sale price that can be sustained by the user in one market fixes the theoretical surplus profits for the promoter, and the difference between this surplus profit and the average profit fixes the tribute that the promoter will agree to pay to the landowner in exchange for access to the site. Thus, it is eventually the use of the land that settles its price.

Thus, we can say that it is not because land prices increase that housing prices do: on the contrary, it is the rising purchasing power of the ruling classes which proceeds to certain neighbourhoods which makes land prices increase.

Finally the administration, confronted by the presence of promoters, owners and users, carries out investment and fixes the land occupancy rules. The urban land tribute per unit of capital determined by the above mechanisms is then transformed into a tribute per unit of area. This is how land prices per square metre are finally fixed.

It is evident that at this point the system 'loops': this outcome can in turn modify the state of the construction system, etc. With regard to the 'edge' problem between two uses of land (in the ESDS), it has to be approached from a dynamic angle, for example by means of logistic curves.

This is the way the land price pattern is determined. Now, we know that this urban tribute is a part of social surplus value. Where does this surplus value come from?

The Sources of Land Tribute

Reading this one can recall Marx's judgement over 'monopoly' rent: 'Urban owners make others pay for the right to inhabit the land'. It is actually the first 'source' of urban rent which Engels worked out: thus we will call it 'Tribute à la Engels'. But there is another one, related to that evoked by Marx in the agricultural case: 'Tribute à la Marx'.

Tribute à la Engels. It is clear that the land tribute's structure gives great importance to the purchasing power of middle and superior classes; it is also necessary to recognise that what is implied in this case is not so much the right to inhabit the land but the right to not cohabit with anybody.

What then is the social relation that is involved? It is the relation between the capitalist class as a whole (with its vassal classes) and urban property. This tribute is actually a redistribution of social surplus value, which shifts first in the form of strongly skewed incomes and then through the selection of places in the social division of space (which is itself very segregated whereas construction prices do not fluctuate much) and is then partially transferred to the parasitic layer of urban property owners.

In Marx's and Engels' time it was mainly this social relation which was affecting housing for all social classes: the middle-bourgeois class of urban landowners could, due to the rural exodus and to the mediocrity of dens, easily extort a 'Tribute à la Engels' from the working class. This would raise the cost of labour power, levying a share of the surplus value extracted in the industrial world as a whole. This secondary contradiction within ruling classes (between 'productive' capitalism and 'parasitic' urban property) turned little by little in capital's favour (1923 and 1948 laws), but still keeps a distinct importance in France, given its archaic social structures.

The 'Tribute à la Engels' particularly, is and will be the principal source, the 'womb' of differential tribute linked to the social division of space. For, contrary to the case of wheat farming where differential rent is directly quantifiable (and this is not so in the case of great vintages!), the difference between houses is qualitative and it only becomes quantifiable through the structure of differentiated purchasing power of different social classes.

The Tribute à la Marx. During the sixties, in France, housing 'producers' tended to become increasingly powerful users of suburban land, thus moving the housing crisis from a quantitative to a qualitative one. Their goal was then to transform housing into a good free from monopoly rent – that is, into a product subject to autonomous feasibility conditions (comparable with the general circulation of social capital). We could say without 'abnormal' transfer of surplus value from one sector to another. But, we have shown that construction is precisely one branch where the ratio of surplus value to productive capital employed exceeds the average ratio. And this is no more of a coincidence than in the agricultural case. Due to its 'pre-industrial' nature, not only is the organic composition of capital in house building, as we have seen, rather low, but also the exploitation rate in this sector is in France (and for connected reasons) higher then elsewhere.

We detect here a new contradiction, this time between the urban owners' interests and that of the 'big builders': the need to set aside a part of the surplus value for the land tribute prevents intensive accumulation of capital and industrialisation in the building sector.

The big firms drew a logical conclusion from this: 'The search for building sites and their first laying-out goes far beyond the builder's duties, from both the financial and legal points of view. It is just as abnormal to ask this builder to provide the plot, as to ask an automobile producer to provide the road. The solution seems to be a formula similar to the present 'ready-to-build' one, where the serviced plot will be furnished by the administration and will be set at the disposal of a producer, who will be in charge of the design and realisation tasks, in the manner of a 'turn-key' project'.[9]

Scope and Limits of the Distinction between the Two Different Sources of Tribute. We have so far voluntarily stuck to the somewhat naïve way in which Marx and Engels presented the problem of the land tribute's 'sources'. As if the created value were a liquid substance, springing up from social labour into different branches of production and distributed afterwards in various revenues. We could then follow the trail of each 'molecule of value' as with a radioactive tracer, from its creation till its absorption . . . Thus, the land tribute 'à la Marx' would distinguish itself by the fact that it would be value created in the construction industry which would fall into the owner's pocket, while in the case of the tribute 'à la Engels' this value could have been produced in any branch.

Actually, the link between the distributed income and created value is more global and more fuzzy. Just as we say in French 'money doesn't have a smell', so circulating value doesn't remember its origins. Since it is all the social labour which is subject to the abstraction of price mechanisms, the part due to the landowner doesn't come more particularly from the capitalist production carried out in his plot, than from any other. The fact that the value created on his plot is superior (as in the case of housing, agriculture) or not (the oil case) to the price of production doesn't change anything. In a sense, every tribute is then a tribute 'à la Engels', a part of total social surplus value granted to landed property.

The laws which determine the level of this tribute are not the same as the ones which determine the surplus value level. Whereas the latter are concealed from the eyes of the economic agents and will only appear through theoretical analysis of the capitalist society's 'anatomy', the former are the subject of negotiations, contracts, public regulations, and even of international treaties.[10] Nevertheless, at a general level, the famous 'Marxian equations' remain perfectly verified no matter what has been said about it (see note 7). If we choose unities such as:

sum of added values = sum of net product prices,
then: sum of surplus value = sum of profits and rents.

If this is so, what is the advantage in distinguishing tribute 'à la Marx' from tribute 'à la Engels'? Actually, behind the dubious hydraulic analogy, those authors were trying to detect where the social existence of landownership makes an impact in the capitalist accumulation cycle. In the case of tribute 'à la Engels' the articulation takes place only at the circulation level. When speaking of tribute 'à la Marx', one tries to point out (as we have sketched above) that the existence of landownership exerts its effect right down to the production process.

We get from this some really different types of antagonism or alliance between the distinct fractions of the industrial and financial bourgeoisie and the various types of land owners.[11] We are able to distinguish (Lipietz 1974) the different types of land reforms that have been proposed in France during the last fifty years, assailing in their different ways both sources of the land tribute, corresponding to distinct social contradictions, seeking for different class alliances.

Modulations and Variations of the Land Tribute

If we have shown how a specific use of land brings about a land tribute, and if we have shown what its sources are, what relations link capital and urban ownership, then we still have to account for the modulation in space of this tribute and its variations with time. (We will only consider here the land tribute brought about by capitalist housing production.)

If in this section, reverting to custom, we keep the term 'differential tribute' to designate the inequality of tribute provoked by the diversity of forms and conditions of the valorisation of capital. Let us simply remember that, for us, this tribute doesn't miraculously 'add' itself to a hypothetical 'absolute tribute' determined somewhere else by another use. Again for simplification purpose, we will use the word 'tribute' to name the surplus profit even when it is not actually paid to the urban owner, as for instance when the promoter 'fools' the owner with regard to the potentialities of his plot.

We can first study the tribute's modulation in space. It arises from the articulation of basic social relations which determine the land tribute's sources in the local differentiation of the practices of the economic agents. These practices are very different if we talk about wheat production or about housing. Hence the vanity of attempts to transfer term-by-term the (already not very specific) Marxian concepts of 'differential rent of fertility', 'situation', etc.

We are driven to a basic distinction between two types of differential tributes: those which are independent of the promoter's autonomous (private) practices (we can call these exogenous) and those which depend on the promoter's practices (endogenous).

Exogenous Differential Tribute

An exogenous differential tribute can exist when the surplus profit on invested capital is not only limited by, but is also determined by conditions due to the site's properties.

In the case of housing a first type of differentiation is introduced by the conditions given to the labour process – that is, by problems encountered by the building process, conditions which sometimes can be 'physical' but generally are social: presence of old diggings, of more or less load-bearing soils, of ancient buildings to be removed (and we can add to this the cost of reaccommodation of former users – cost in money and in wasted time on immobilised capital).

We will name this type of differentiation: differential tribute of constructability. It modulates surplus profit in terms of cost price.

A second type of differentiation (and it is the most important) is obviously the economic and social division of space. From the moment it is inscribed, drawn out over the map, even as a 'project', every single limited operation (that is, one which does not modify this ESDS) is under the obligation to pay to landed property at least the tribute brought about by the locally-prevailing use: and it can do it because it is precisely the purchasing power of the social category of final users which is the source of this type of tribute. Otherwise the tribute linked to situation in the ESDS is, by its nature, by the underlying social relations, a tribute 'à la Engels'.

We can notice here the removal of the ambiguity introduced by the use of expressions such as 'tribute brought about by such a use of land'. It is not the promoter's fancy that, by his choice, determines the price of ground. It is the social division of space, globally determined, at the scale of the whole social formation, from the city's past. In the medium-term functioning of the development system, the ESDS virtually imposes itself on the promoter. It dictates to him which valorisation process will produce the possibility of such a land tribute.

The mechanism of differential tribute of social situation is precisely the economic process (though there exist others) which stabilises the ESDS, by assuring the suitability of the housing produced to the social status of the neighbourhood.

It remains for the objective basis of the ESDS (quality of streets, of general architecture, illumination, transport services, neighbourhood services) to be produced by human labour, by a massive investment of private or public capital. Thus we approach the 'endogenous differential tribute' problem.

Endogenous Differential Tribute

We will not dwell for long over the 'extensive endogenous differential tribute' by which, over two equal plots 'equally located', a promoter will agree to pay a tribute twice as high if he can produce and sell twice as many houses.

We shall insist nevertheless on this point: the regulation of the 'Ground Occupancy Coefficient' (GOC) intervenes as a limit to the promoter's autonomy, but the tribute is actually brought about by the action of the capitalist promoter himself. Yet once the legal GOC is known, the landowner will naturally demand the maximum tribute consistent with the GOC and the situation in the ESDS.[12]

We will persist longer over the 'intensive differential tribute', that is, where the surplus profit rate depends on the level of invested capital. As a matter of fact, this is more or less always the case, but the concept's relevance is particularly clear in the case of a massive change in the use of ground – for example, from agricultural use in housing (urbanisation), or making an 'in' neighbourhood out of a 'lumpen' one (renovation).

Let us take the urbanisation case. The agricultural capital implanted in the ground is no longer considered of any value. It is necessary to invest first in the primary roads and various networks (water, electricity, etc.), then in the secondary networks, then in the first buildings. The cash flow is at first nil, then fast increasing, then decreasing with the number of floors; such is the rate of return on invested capital. Figure 7.1 allows us to compare the return to the cost price and to the average profit (of rate e).

We can see that the rate of profit on marginal outlay is a strong function of the global level of investment. Hence the level of tribute depends on the mass of invested capital.

This is the reason why a highly sophisticated set of forms of public regulation and financing has been set up in France (for instance with responsibilities for the financing of the primary transport network), in

Figure 7.1: Return with Increments of Capital Invested

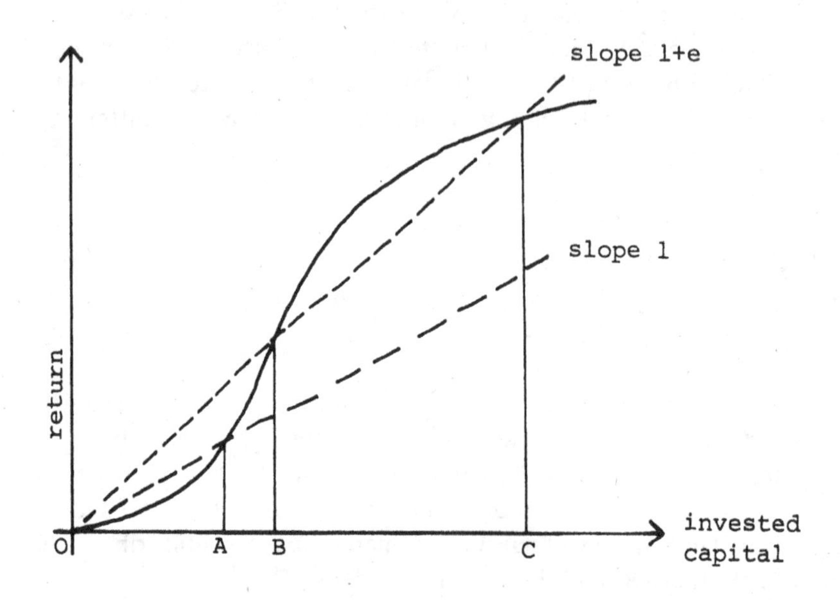

OA: heavy investments are made but there are few inhabitants to pay for them. Only public funds can take up this stage of urbanisation.
AB: houses sold can cover their costs, but the density is too low to yield an average profit.
BC: private capital makes a surplus profit.

order to give private building capitals opportunities to invest directly at the profitable stage (BC in Figure 7.1). Other regulations have been made in order to prohibit landowners from appropriating on their sites the intensive endogenous land tribute corresponding to these huge investments of public or private capitals.[13]

Variation of Land Tribute through Time

Variations show two components: a general element (average price movements of urban land) and a local one. After World War II these movements were generally upwards. This allows a commercial game, not purely speculative, which consists in buying for reselling after the rise, pocketing a 'land surplus value'.

General Variation of Land Tribute. The land tribute can grow because the total profit increases in the production related to the ground, or because the average profit rate decreases. This second cause expresses the saturation of the system of expanded circulation of capital, due to a tendential decline of the profit rate, to overproduction crises, or to other factors (e.g. return of colonial capitals). The first is essentially due to an increase in the tribute 'à la Marx' (industrialisation). On the other hand, urban development immediately makes a differential tribute of social situation appear over all plots previously urbanised, financed by the middle classes forced to pay higher prices for centrality (tribute 'à la Engels').

Finally, let us re-emphasise what may be obvious: that the incorporation of capital (particularly public capital) in land increases its price just as much as density-increases do.

Local Variation of Land Tribute. By definition, this is a question of differential tribute. It can be due to: (1) a variation of extensive endogenous tribute, limited by variation of the legal GOC, for example, the removal of a construction ban; (2) a variation in social use, which can be the consequence of an action by the promoter himself, of some public development (creation of a primary network) or of a move in 'l'air du temps' (e.g. the lofts in New York). According to the case, we can talk of intensive endogenous tribute variations (variation of marginal surplus profit realisable from the housing once the infrastructure is modified) or we can talk of an exogenous variation in the ESDS.

The So-called Land Surplus Value. We have assumed that surplus profit falls into the urban landowner's pocket. Obviously this is not always so:

(1) A speculator (etymology: someone who waits and observes) can buy a plot according to the land tribute corresponding to the initial use (e.g. agriculture) and resell it to a promoter at a price including the subsequent local tribute.

(2) A promoter can carry out the same operation and make a surplus profit which doesn't completely transform itself into land tribute.

These two operations are not equivalent. Both are allowed only by the landowner's ignorance concerning the plot's possibilities. Nevertheless the pure speculator only 'waits' for society to produce (in his plot's vicinity) the increase in land tribute, while the promoter is, at least partly, this society's direct agent, changing the use of ground by his investment.

We use the term 'land surplus value' for the difference between the plot's purchase price and the land tribute corresponding to the final use. This 'surplus value' is not related to the Marxist meaning of the term; it does not correspond to any 'added value'. But the term is prevalent enough (in French) for us to adopt it. Surplus value can be the object, on the part of an intermediate owner, of an active appropriation (if it brings about, by his investments, a local increase in the land tribute) or of a passive one (if he only takes advantage of the local or general variation of the land tribute).

Conclusion

What do we get from such a theory? What do we get from this heavy conceptual apparatus?

First a representation of reality matchlessly finer, a capacity of integration and explanation of facts and figures without measure from Mayer's or Alonso's models. (The epistemological status of those models is a quite separate weakness.) But is the theory 'operational'? It will be necessary first to know what that means. Is it a question of having a model which allows us to calculate, from some features of a plot, its average price? We can first note that Alonso-Mayer models cannot do it. Anyway, it is clear that a theory cannot act directly as an operational model in economics any more than in mechanics. But the theory gives us the components of the model which remains to be constructed:

(1) In a general way, land price depends on the average profit rate and on the purchasing power of social classes, on the pugnacity of landowners . . .

(2) Locally it depends on the 'residential quality', on the construction costs, on the permitted land use and density (GOC) . . .

Essentially the theory shows us in which direction this or that exogenous action or change will make the land price move and how that movement interacts with current social practices and patterns. The theory can then be interpreted in two different ways.

From Sirius's viewpoint, which is that of the academic research inspiring this paper, it allows us to interpret the evolution of the French bourgeoisie's politics in a surprisingly fine way.[14]

From the normative viewpoint it will allow us, we hope, to avoid the loose debates over the respective merits of those various fiscal, regulatory or financial measures which, by only intervening on a unique type of tribute, cannot resolve the urban problem by themselves.

Acknowledgement

Thanks to Ricardo Hausmann for his help with the translation, but I assume responsibility for the result.

Notes

1. The full development of the issues presented in this paper are to be found in Lipietz (1974).

2. This 'tolerance' was far from being complete. The first publication of the present theory had to appear under a false name (Juillet 1971).

3. Applications to the case of oil rent can be found in Lipietz (1974), and especially in Hausmann (1981).

4. These analyses are to be found in *Capital*, Volume III, and *Theories of Surplus-Value* (about Ricardo). Both of these books remained as drafts when Marx died, and many contradictions remain between these texts (see Chapter 9 – eds.)

5. In the seventies several French scholars (e.g. Aglietta 1979; Coriat 1979 and Lipietz 1979a) emphasised the specificity of this regime of accumulation, and called it 'Fordism' (following an old insight of Gramsci). The role of housing in post-war French Fordism is very peculiar: though the new standard of housing was necessary for Fordism to develop, house building itself failed to participate in the trend of productivity borne along by Fordist processes of production. On the contrasted logics of motor-car and building industries, see Lafont, Leborgne and Lipietz (1980).

6. This oddness of a part of the surplus-product which appears as an element of production costs confused even Sraffa (1960). Hence the inconsistencies of his chapter on rent, noted by several scholars. For an analysis of these inconsistencies, see Lipietz (1980b).

7. There has been a lot of arguing about the results of this transformation, because the 'standard' solution of the problem (Seton-Okishio-Morishima) exhibited some paradoxical results. A closer examination of this standard solution plus the discovery of a new solution by Demunil, Foley and Lipietz made clear the following (see Lipietz 1982). Everything depends on the definition which is adopted for the value of labour power. If it is the 'value of the commodities purchased through wages', then the value of the commodities purchased through profits equals the total surplus value.

If, instead, we take the value of labour power to be the portion of the value added which is assigned to wages, then the sum of profits equals the sum of surplus value. (These sums relate only to the net product.) This last theorem has been extended to the case of fixed capital and rent in Lipietz (1979b).

8. Anyway, the word 'monopoly' pervades Marx's texts about rent (see *Capital, III*). Landed property, as a social relation, is from the beginning a 'monopoly on defined parts of the earth' (p. 8), just as a capitalist class could be defined

by a monopoly on the means of production. So any land rent is a monopoly rent.

Even in the case where 'absolute rent + capitalist profit ≤ surplus value', the price of the product is not defined only by the condition of capitalist production, but also by a bargain between the user, willing and able to pay for a plot, and the landowner, able to levy a tribute or to forbid the use of the plot, no competition of capitals being able to raise this barrier. So this price is also a monopoly price, as Marx admits (p. 146: 'But, whenever the absolute rent would be equal to the totality or only to a part of the excess (of value over price of production), the price of the agricultural product would be a monopoly price, because it would be above the production price'.

Moreover, 'though landed property is able to bring the price above the production price, it is not itself, but the market conditions, that will define up to what point the market price will approach the value and by-pass the production price'.

So the existence of landed property is the cause of the rent, but not the determinant of its magnitude. The quantity 'value minus price of production' is of no relevance at all. There is no difference in Marx between 'absolute rent' and 'monopoly rent', either as a concept or as a practical mechanism. The only difference could be perceived in national accounting! Yet there is an idea in this distinction, that we shall catch as 'tribute à la Marx'.

9. That was the answer of M. Pagezy, top manager at St Gobain-Pont-à-Mousson, to an inquiry on the industrialisation of building activity.

10. On the distinction between these two kinds of laws, see Hausmann and Lipietz (1983).

11. For instance, as far as tribute is 'à la Engels', there is a community of interests between landowners and capitalists investing on their plots, against the rest of the society. This was quite noticeable in the early seventies between oil companies and OPEC. As far as it is 'à la Marx', the contradiction splits between landed property and intensive accumulation in land-using sectors. Hausmann (1981) suggests a distinction within the landowner-user relation: the relation of access and the relation of payment. The first is more relevant to the effect of landownership on the forms of accumulation, thus on the tribute 'à la Marx'.

12. As Kascynski (1982) recently pointed out, one peculiar form of this tribute is the parcelling-out of vast pieces of field suitable for 'rurbanisation' (individual house-building in the countryside for urban workers), at present the most important form of house building in France. Using the present theory of land tribute, Kascynski was able to construct a model and a methodology of land prices observation in the region Nord-Pas-de-Calais.

13. One may notice that in essence the difference between 'exogenous tribute' and 'endogenous tribute' (at least, as far as the intensive form is concerned) refers to the fact that, in the first case, the investment is done in a pre-given social framework, without modifying it, and in the second case it changes it. This distinction is connected to another: 'competitive regulation vs. monopoly regulation' of the production of social space (Lipietz 1980a).

14. In Lipietz (1974) last chapter.

References

Aglietta, M. (1979) *A Theory of Capitalist Regulation – The US Experience*, New Left Books, London

Alonso, W. (1964) *Location and Land Use*, Harvard University Press, Cambridge (Mass)

Bettelheim, C. (1975) *Economic Calculation and Forms of Property*, Monthly Review Press, New York and London

Coriat, B. (1979) *L'atelier et le chronomètre*, Bourgois, Paris
Dutailly, J.C. (1970) *Les valeurs foncières en région parisienne*, I.A.U.R.P., Paris, mimeo
Engels, F. (1969) *La question du logement*, Editions Sociales, Paris
Hausmann, R. (1981) *Oil rent and accumulation in the Venezuelan Economy*, PhD Thesis, Cornell University
—— and A. Lipietz (1983) 'Esoteric vs Exoteric Laws: the forgotten dialects – Marx on the divergence between value-production and nominal revenues', *New Left Review*, January
Juillet, A. (1971) 'Sur la rente foncière urbaine', *La vie urbaine*, no. 4
Kascynski, M. (1982) *Observation foncière et division économique et sociale de l'espace. Le cas de l'agglomération d'Arras*, Thesis, Lille I
Lafont, J., D. Leborgne and A. Lipietz (1980) 'Redéploiement industriel et espace économique', *Travaux et Recherches de Prospective*, no. 85
Lipietz, A. (1974) *Le tribut foncier urbain*, Maspéro, Paris
—— (1977) *Le capital et son espace*, Maspéro, Paris
—— (1979a) *Crise et inflation: pourqui?*, Maspéro, Paris
—— (1979b) 'Nouvelle solution au problème de la transformation: le cas du capital fixe et de la rente, *Recherches Economiques de Louvain, 45*, December
—— (1979c) 'Les mystères de la rente absolue. Commentaire sur les incohérences d'un texte de Sraffa', *Cahiers d'Economie Politique*, no. 5
—— (1980) 'The structuration of space, the problem of land, and spatial policy' in J. Carney, R. Hudson and J. Lewis (eds.), *Regions in Crisis*, Croom Helm, London
—— (1982) 'The "So-Called Transformation Problem" revisited', *Journal of Economic Theory*, no. 1, January
Maarek, J. (1964) *Recherche sur l'urbanisation spontanée*, mimeo, S.E.M.A., Paris
Marx, K. (1960) *Le Capital, III*, Editions Sociales, Paris
Mayer, R. (1965) 'Prix du sol et prix du temps. Essai sur la formation des prix fonciers', *Bulletin du P.C.M.*, no. 10, Paris
Sraffa, P. (1960) *Production of Commodities by means of Commodities*, Cambridge University Press, Cambridge
Topalov, C. (1970) *Les promoteurs immobiliers*, Centre de Sociologie Urbaine, Paris
—— (1973) *Capital et propriété foncière*, Centre de Sociologie Urbaine, Paris
Vieille, P. (1970) *Marché des terrains et société urbaine*, Anthropos, Paris
Wingo, L. (1961) 'An economic model for the utilisation of urban land', *R.S.A. Papers and Proceedings, 7*

8 CAPITALIST URBAN RENT

Ambroise Gravejat

Building land prices are usually explained by two theories: situation and monopoly. The former stresses cost of transport, external economies of the conurbation and public equipment. The second theory admits the influence of law and social organisation which gives the landowner almost absolute power over land (Derycke 1979; Granelle 1970; Guigon 1982; Guyot n.d.). The two explanations concur fairly well with prevailing economic theories (Ricardo, Marx and neoclassical theories) and they are certainly borne out in most analyses of human groups both in space and in time.

In France, the development of Marxist theory from 1965 onwards enabled criticism first of the empirical and factorial studies of urban development, by giving a general and coherent point of view of social organisation as well as urbanisation. In particular we may quote the studies of Henri Lefebvre (1968; 1974), Manuel Castells (1972), Christian Topalov (1972; 1973) and Alain Lipietz (1974). These studies mainly emphasise the relationship between social and economic formation, urbanisation and the role of differential rent II in the analysis of urban rent.

With this problem in mind, we have essentially attempted to link land price to building costs and to the general profitability of the economic system. The notion of building rents underlies an explanation of urban ground rent. This paper presents the elements of the method, a synthetic view of various rent systems and of the specificity of capitalist urban rent. Attention must be drawn to the fact that the study first aimed at finding a simple mechanism to explain building land price based on these problematics. The explanation could reveal 'mechanisms' such as those which exist for differential rent I or differential rent II. Analysis of urbanisation as a whole is not the aim of this paper.

Building: A Necessary Mediation for Urbanisation

A number of points of view and elements of method became apparent as the study progressed.

Figure 8.1: Social Formation and Urban Form

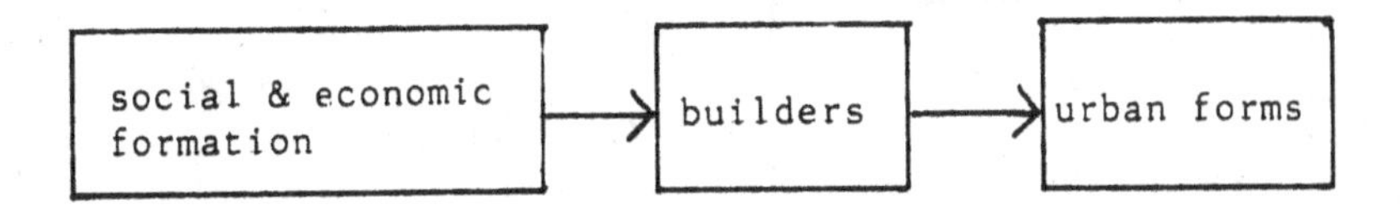

From the Mode of Production to the Producers of Space

It appeared clearer to study the agents which intervene in the production of space, rather than begin from the analysis of the means of production as a whole and its relationship to space, that is, the role of local authorities and builders (Goux 1978). From the beginning of the nineteenth century, the building and public works sector has seemed particularly independent of political power and apparently obedient to prevailing economic forces. This has justified an economic approach dissociated from political analysis. This mediation implies the possible existence of distortions between urbanisation (which ought to be adapted to the urban social formation) and what is actually carried out by builders. This is shown in Figure 8.1.

This point of view does not correspond exactly to the usual distinction in economics between production and consumption. In fact, demand includes the factors generating urban extension: population, density, concentration, economic activities, means of transport, income levels. These various factors are presumed known for a given town at a stated period. They are considered as necessary elements for urbanisation, though insufficient by themselves.

The building sector itself will first be analysed in its productive role at a given period: the state of building techniques, labour organisation, financial conditions and allocation of land space. These elements define its production costs and, interacting with the ability-to-pay of potential users, they explain the decision to build. All these factors explain the evolution of building but not the existence or the level of urban rent. This leads to a second simplified diagram, shown in Figure 8.2.

Figure 8.2 shows the outlines of causality. Each rectangle assumes the existence of characteristics which are specific to a given area and which create distortions or incidental or structural oppositions, hence explaining the appearance of conflicts or contradictions in the 'functioning' of the whole system.

Figure 8.2: Social Formation, Agents and Urban Form

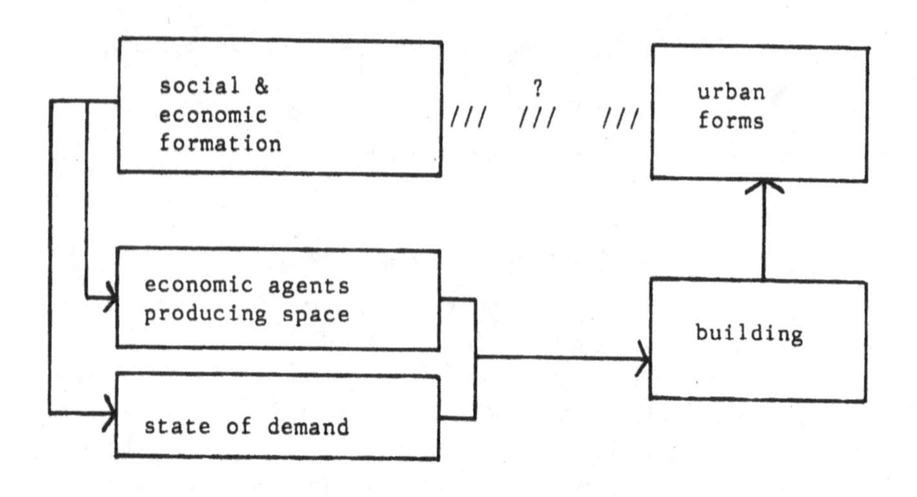

The Landowner

Until now landowners have been ignored. They appear to play no part in the decision to build. This first remark is important, as not every change in demand has any direct effect on property. Its primary effect is on potential builders. Thus we can see that any change in activity, population, environment, infrastructure or transport will encourage the promoter to build and develop the area as much as possible. Logically he will seek a normal return on the capital invested. Consequently, the final quantity of profit will depend on the quantity of money capital put in.

According to classical assumptions, the landowner does not intervene directly in the process of production. He owns a property which gives him a known income and the investment represented by the building will make him the owner of goods which, once paid for, will give him a higher income. Figure 8.3 demonstrates the modification of the landowner's income in time.

Thus the landowner is generally receiving agricultural rents until time t_1. In this example, building takes two years during which time the owner receives no income. Then he has to repay the building cost with the rent he collects, over a period of ten years to t_3 – as shown in Figure 8.3. From this point on his only expenses are maintenance and taxes. This simple diagram is interesting because it clearly demonstrates normal development of urban property. It will be altered if we

Figure 8.3: The Landlord's Income through Time

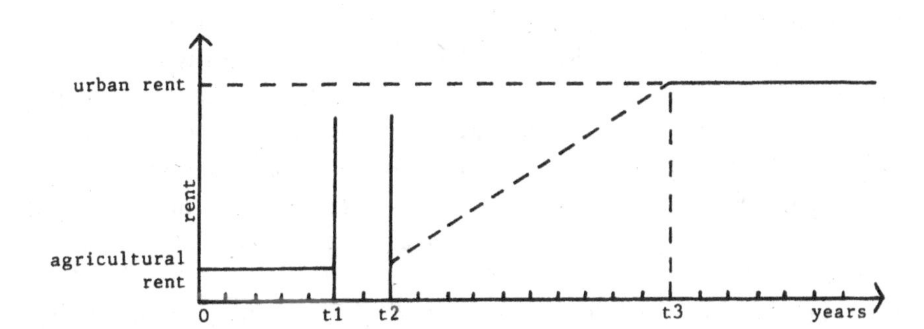

suppose that the land was already built on at time t_1 but that the new operation causes an intensification of building – increased height, additions or internal intensification. The notion of rented property is thus defined by four factors: (i) the right to ownership of an asset, (ii) the non-reproducibility of that asset, (iii) the usefulness of the asset for the reproduction of society and (iv) the renting out of the asset.

We find ourselves back with Ricardian theory, with two exceptions. The asset is partially produced, difficult to reproduce, and it remains lucrative after repayment of the initial investment (this expression, 'avances primitives', is borrowed from Quesnay). This modification corresponds to the case of Marx's analysis of differential rent II. Nevertheless, the facility to go on collecting rent is not linked to the land, but rather to the building, hence the value of using Quesnay's expression.

A second difference remains. Does this continuing rent flow represent anything productive? Certainly not. Housing is a product like other goods and the rent is simply a long-term payment. The building cost is paid back within a fairly short period of time and the financial return which the owner ultimately derives from it can then be defined as rent. The rented property is thus defined in a social relationship which allows an individual to appropriate someone else's income without working. We must not confuse the case of an owner occupier, who uses an asset without controlling anyone else's income, with the case of the owner who rents out his property.

From the Individual Agent to the Function of the Agent

On the basis of the preceding outline, we may now define the status and economic role of any urban landlord; in other words we can define the 'function' of urban landlords or the place which they occupy in the production and management of urban development areas. It is difficult to use the term 'function' today because it is so closely linked to the functionalist movement which emphasises capacity to satisfy primary or derived needs in order to define a function. But the term 'element of structure' is no better. So we have to make do with the expression 'place' or 'role', terms from geography or the stage, which are hardly satisfactory. We ought in fact to speak of 'function' with the meaning of 'organ' or 'organic position' in a social system. Because of his legal position (right of ownership), the owner has the function of exacting maximum profit from a property and therefore has to assemble there the development most likely to provide this income. This is the development process for urban land.

This role may be played by numerous agents in various processes. For example, the property developer who hides the intended and use of the land he buys, thereby paying less than the real market price, appropriates a part of this rent for himself. Similarly, the landowner who sells land for building in fact capitalises future rents. The diversity of agents able to appropriate these rents and the many processes of transforming space do not alter the real nature of the rent phenomenon as defined. The forms of its creation and appropriation may change, but the general definition remains valid. Two main forms of appropriation exist in the present system: ground rent and housing rent. There are several forms of production of these rents: building, restoration, modification of the environment. This third factor would appear distinct because of the direct relationship between public or private transformation of space and the constructor's activity. However, the changes in environment may be too slight to justify a new building and in this case a portion of the advantage would be appropriated by the private owner. The same would be partly true during the whole period preceding the new building. But the influence of this factor will be limited by the form and thus by the present use of the property.

Rent Systems

Although it may be possible to define a general rent concept derived from the value of labour power and the non-reproducibility of certain monopolisable assets (nevertheless useful for the traditional reproduction of society), it is apparently impossible to define its constitution, its level and the different forms of its appropriation independently of the analysis of specific urban systems which are related to the dominant economic and social system. We shall discuss two of them which demonstrate the specificity of capitalist urban rent.

The City, Rental Property and Rural Production

Notions of rent and property immediately call to mind the feudal system or Roman villa. However, we should not reject this association as being something of the past, nor should we confuse all rent systems together. In fact each economic and social system is defined first by a system of ownership, both of the land and of the means of production, and then by the supportive production relationships which sustain it. Similarly, specific rent systems adapted to the dominant production method develop in the towns.

In Rome at the end of the Republic and the beginning of the Empire, management of rented urban buildings was mainly based on economic logic, so this approach is therefore right and possible.

Rome: a City of Tenants

Leon Homo (1951), like Jerome Carcopino (1939), describes a city in which 50 per cent of the inhabitants lived in uncomfortable and poorly maintained accommodation and where complaints about the cost of rent were regular. This state of affairs did not improve as the Empire grew. Rich families had large houses built in the urban outskirts, but most of the population lived in accommodation without either heating or water. The Emperors had to issue edicts ordering the construction of stone staircases to prevent the risk of fire and the doubling of brick walls in order to strengthen the buildings. It is difficult to imagine that the Romans, among the most prolific builders of antiquity whose works have in some cases lasted through to the present day, failed to build healthy and comfortable ordinary dwellings for themselves. When we look at Roman remains and archeological information, it is obvious that the state of the art and the know-how of the building trade could not be blamed for this shortcoming.

High Yields on Urban Property. We do not possess accurate data on building costs, building craftsmen's wages or effective purchasing power. Tenants' complaints about rents might be just as indicative of quarrels with rich families as of low incomes. However, some facts cannot be questioned:

(1) The low cost of building: although aristocratic families made use

of the know-how of the members of guilds (each trade had its guild) for the construction of public buildings, they only needed slaves for building and maintaining accommodation for rent.

(2) Poor maintenance: the tenants did not complain about absence of water or heating facilities, but rather about cracks in the walls, water seeping in from the roof, broken-down shutters and even staircases, and the excessive congestion of living space – two or three rooms, or the cellar, attic or even the space under the stairs.

(3) Speculation on these buildings: management practices that tended to rent unfit buildings as dwellings. For example, Crassus owed much of his wealth to real estate and boasted 'that he had never had a single dwelling erected himself for it was much more profitable to buy tumble down housing at low cost for renting out after minimal repairs' (quoted in Mumford 1964).

(4) The high cost of building land: this was the major cost at the time. Building costs properly speaking (work of slaves and craftsmen) were not mentioned in records, perhaps because the figures were unknown because they came under private management and also because they amounted to less than the cost of the land. Contemporary writers mention the cost price of land for official public works but do not mention other costs. The only known case is that of Cicero's house. Leon Homo calculated that the price of the land represented 43 per cent of the total cost of the house. But we must add that it was a sumptuous 'domus' using expensive materials and the services of skilled labour. Based on a similar figure for cost of land for a rented building constructed by slaves in ordinary materials, it is quite clear that land must have exceeded 60 per cent of total building costs.

(5) The high returns on these buildings: Aulu-Gelle writes: 'the properties located in town bring in high profits but they also carry a high risk. If it were possible to prevent those catastrophes to which Roman houses so often fell prey, I would lose no time in selling my fields to become an owner in the city' (Homo 1951: 45). The catastrophes referred to were fire and collapse. We may therefore deduce that urban returns were higher than agricultural return, which itself was quite substantial (see Cato's analyses).

These facts lead us to conclude that rents were fixed without any direct calculation against building costs (cf. slave labour, ordinary building materials, but high land cost and high rent). The high returns which considerably exceeded the repayment of costs must come into the category of rent. They are associated with ownership and not with work.

The Attractions of the City and Urban Advantages: Citizenship and Comfort. How can these high rents be explained? Building and maintenance costs can immedately be ignored because they were minimal. The price of the land itself was not an explanation. On the contrary, it was another form of expression of the high income from urban property. The price of the land is always considered as capitalised property rent, the notion of monopoly, in the ordinary sense of the term, does not stand up to analysis any better. It would not allow the establishment of a mean level of rent nor demonstrate any specific economic relationship. Above all, any citizen who wished to escape such monopoly would have had the opportunity of living on his rural properties. In other words, what were the advantages of urban living during this period?

First of all, every Roman citizen had to live in Rome if he wanted to take part in government (direct representation) and so best manage his business, that is, his rural properties. Laws were drawn up in Rome; military defence of the territory was organised there too. Citizenship and social and military affairs obliged free men to live in their city, which was considered the only forum of political and civic expression. The term 'Freedom of the City' sums up these various advantages very well.

During the Empire the city itself offered a whole host of material advantages. Thus the great patrons built not only the temples and basilicas needed for communal life, but also sanitation (aqueducts, fountains, public baths, thermal baths, sewers and public latrines), buildings for leisure and culture (circuses, libraries, theatres, amphitheatres, naumachia). Urban residence enabled a man to benefit from all the advantages of civilisation, whether in arts and letters or just domestic life. All this building was financed by political candidates for the highest offices and therefore in the long term by the incomes they drew from their rural property, the management of the provinces or their conquests (Gravejat 1980).

The Roman resident did not have to pay for any of these rights and services. Thus rent could be considered as the synthesis and monetary expression of these urban advantages. The landlord therefore benefited from the social and political services provided by the great patrons, because of the rents he was able to charge. This makes it easy to understand why senators and noblemen, as well as the middle classes, wanted games and feasts. A rich and powerful candidate was able to entertain the people while enriching his peers at the same time.

So who actually paid for this rent? If it functioned according to the theory described here, the rent would in fact pay building and maintenance costs (minimal) and for all of the services offered by the city (citizenship, comfort and games). Although these were not financed by the owners, they existed nevertheless, and a real service was provided in return for rent in this case. This is a long way from monopoly rent in the crude sense of the term.

However, the counterpart to right of citizenship was rural production by slave labour, and the distinction was very clear here: the country produced and the city consumed. So a great part of the rural slave's product was transferred from the rural landlord to the urban landlord in the form of rent, while the latter, in this case, gave nothing in return. This first situation corresponds to monopoly rent in the economic sense of the word (transfer of surplus product from one branch to another without any real compensation).

The situation is similar in the second case, with the addition of imperialism. The rent obtained by the urban landlord reflected urban advantages (social and entertainment facilities) which in the final analysis were financed by the 'surplus labour' or the 'net product' of rural slavery or by the tribute paid by conquered provinces, which itself is nothing more than 'net product' in political form. A part of the military or economic gains of the few were redistributed among all the owners in the form of rent. Slavery and imperialism were the two forms of exploitation which supplied the city economically and ensured its survival. It is easy to understand that the urban tenant was a privileged person in this system. Even if he protested against the high rents, he could not query an economic and social organisation which allowed him to be one of the political and urban privileged class. Rent was nothing more nor less than a kind of social redistribution of the incomes of a few important families within the urban society. Slavery and imperialism maintained and supported tenants who themselves paid rent to a mass of owners. But the economic basis of the whole system was still the agricultural surplus labour of slaves in the country.

Free Cities, Trade and Mansions

Trading cities developed in the West from the twelfth century onwards and their great era of prosperity occurred between the fourteenth and seventeenth centuries. Trading cities and free cities, totally independent as in Italy in the fifteenth century, or protected by powerful monarchs as in Lyon in the sixteenth century, organised democratic and self-governing city administrations in which the activities of tradesmen and bankers, as well as those of the clergy and lawyers, were intermingled.

First these tradesmen exchanged rural products and their earnings resulted essentially from the difference of productivities and product between regions or between nations. Lending at interest, which was strictly forbidden by the Church, was able to develop only on the basis of currency exchange. The exchange rate always included an interest rate, according to Schnaffer's analyses (1957). Later on, 'workers' made their appearance as well as proper workshops. However, this was always a later feature, merely complementary to rural production – e.g. finishing, preparing and manufacturing of goods. Nevertheless, there were a few specifically urban activities – e.g. silk manufacturing and printing (cf. Lyon in the middle of the sixteenth century) (Gascon 1971).

Houses and Mansions. The Roman *domus* had totally disappeared and most wealthy families lived in two- or three-storeyed mansions, while buildings for rent were also being erected. This was the case of the Bourgneuf area in Lyon. In this city in 1550, 2,277 landlords owned 3,561 houses for an estimated population of at least 60,000. So this trading city was primarily a city of tenants. Did the payments correspond to a rental income or did they only represent deferred payment for the building?

A Method which Eliminates Rent. Most studies on the profitability of buildings for rent in the sixteenth century reveal a rate of return between 5 and 13 per cent. Thus Marc Masson (1980) calculates that a day-labourer could spend an annual 10 per cent (6 *livres* and 10 *sous*) of his income on rent. The houses comprised six to ten bedrooms and brought in an average of 10 to 65 *livres* per year gross rent. The cost of a building for the same period is estimated at 500 or 600 *livres*, making a gross rate of return of between 6.15 per cent and 13 per cent. The same calculation can be made from a comparison between the sale price of a building and the annual return on it which gives a gross return of 5 per cent. Gascon (1971) finds rates ranging between 4 per cent and 13 per cent.

Returns and Rents. It is obviously impossible to calculate a rent yield using this method. What in fact is purchased by the real estate buyer? Construction costs or a perpetual annuity? Bernard Schnaffer's studies on the various rent levels in the sixteenth century (1957) clearly demonstrate that real estate income is regarded as rent, in the same way as agricultural income, and that urban property offers the same guarantees of durability and security as rural property. This range of

5 to 13 per cent can be explained in that building annuities could be purchased at 18 deniers, the same as for agricultural land, giving a yield of 5.55 per cent, but they brought in at least as much as money annuities at 10 deniers: a gross yield of 10 per cent. The advantage of this calculation would be to equalise the seller's and the buyer's incomes, thus admitting a minimum rate of return of at least 5 per cent.

This method is more accurate, but it conceals the real origin of rent, which is work and construction costs.

Hugues Neveux's studies (1975) on the management of blocks of rented accommodation in Cambrai in this period give a good idea of construction costs and the time needed to pay for the building through rent. He quotes the example of the Principal Trades Department which asked the Building Department to erect two houses and allowed them to collect the rent for a period of 28 years.

Neveux concludes from this that the house was paid for within about 30 years, in other words that the investment was repaid. The analysis is good, but its conclusion is wrong in our opinion because the Building Department would only have recovered its capital while the Principal Trades Department would have gained two new buildings free. Payment ought thus to have been made over a period of 14 years if the Building Department wished to obtain a profit similar to the Principal Trades Department. If it wanted to add an interest rate on its capital, the payment should have been made within less than ten years. As we shall see, the cost of construction should have been repaid within seven or eight years in order for substantial and long-lasting building rents to be obtained. These buildings have been rented out for 300 years. High figures for rent emerge from this method of analysis. How can they be justified?

Trade Fairs and Mansions. We can immediately ignore monopoly rent in its crude meaning, for it explains nothing. Similarly, construction did not yet reflect fully-capitalist forms and its costs are hard to interpret. Foodstuffs were not bought at their real production cost, and urban labour did not produce on a strictly capitalist basis even though a wage-earning class had already emerged. Finally, the attraction of the city cannot be understood just from an examination of its limited facilities: churches, hospitals and public streets. The city was maintained through taxes paid by the merchants and wealthy classes; public expenditure was hardly excessive.

One explanation remains therefore: the trade fairs and the sudden influx of crowds of people during these periods. A fair lasted 15 days in

Lyon, giving rise to intense activity four times a year. Hotels patronised by the merchants had thus to be paid for in a short period of time. For instance, Florentine bankers used to pay rents of 100 *livres* a year, in a building worth 400 or 500 *livres*. The majority of traders and wealthy foreigners therefore bought mansions or had them built. The outlay was covered within four or five years. The cost of rooms during the fairs must have had a direct effect on rents and low wage-earners and craftsmen, attracted by the hope of work, must have had to pay for their accommodation accordingly. The advantage of this explanation is that it demonstrates the peculiarities of a trading city in comparison with the political city of ancient times or the modern capitalist city.

Finally, what did the tenant pay in addition to construction costs? The merchant drew profit from the useful aspects of urban agglomeration (that is, the simultaneous presence of many other merchants) and the artisan found work.

What kind of excess labour supplied these rents? In the last analysis, if the merchant drew his income from differences of productivity in space, it follows that agricultural workers provided the main source of this rent, plus a few urban wage-earners. We are thus back to a monopoly rent in the economic meaning, in so far as rent meant redistribution of the excess labour of farmers via the traders and bankers into the pockets of urban property owners. Commercial urban rent was thus related to the main urban activity and to prevailing feudal methods of production.

Capitalist Urban Rent

The presentation of these other two systems of rent helps us to understand the specificity of capitalist urban rent. Any theory in which the price of the land is nothing but capitalised rent remains very general. As we have just seen, it applies to all systems of rent. Ricardo's theories on differential rent I (1970: 45) and Marx's theories on differential rent II (1973: Vol. III) now appear in all their originality and the value of their contribution is clear. At the same time, most formal research on urban rent remains at an early stage and so far it affords no real understanding of the phenomena involved.

The Bourgeois Apartment Building and the Working-class Tenement

During the second half of the nineteenth century, urbanisation developed

Table 8.1: Proportions of Rented Accommodation in Lyon in 1869 and in 1911 (% of all accommodation)

Years	District						Whole city
	1	2	3	4	5	6	
1869	66.22	60.72	57.12	42.17	61.55	50.08	56.47
1911	82.46	73.32	78.68	64.44	54.44	76.22	73.24

Source: Aboucaya (1963), Leon (n.d.)

on an effectively capitalist basis with industrialisation and the emergence of joint-stock companies. Production took place mainly in the cities while, in the country, farmers demanded normal rates of return on their capital, even though in France agricultural production was centred mainly on family units.

Urban property owners prevailed. In 1911, in Lyon, landlords possessed 73 per cent of all accommodation (see Table 8.1). During the century, the building trade expanded considerably; the number of houses in Lyon grew from 3,600 at the end of the eighteenth century to 15,790 at the end of the nineteenth century. The number of inhabitants grew from 140,000 to 406,000 (Marmy and Quesnoy 1866; Garden 1975).

Yields on this property ranged from 7 per cent to 17 per cent per year. This range did not stem from the kind of transaction, as had been the case in the sixteenth century, but from the different categories of construction. The long-lasting high quality stone residential building, erected on its own land, yielded an average of 7 per cent to 8 per cent. Public buildings, erected in *pisé* (clay and stone mixture) or even in wood, on plots of land belonging to the Lyon hospital administration, with a 12 year lease, actually yielded as much as 17 per cent (Daumard 1965). This difference clearly shows the importance of the durability of the construction. Working-class housing had to be paid off within a very short period and the investor was not sure whether he would be able to do the same business again. The owner of bourgeois apartments forecast a regular income for several generations. For Paris, Michel Lescure (1977: 428) found gross rates of return ranging between 5.29 per cent and 7.77 per cent in 1901, and between 5.49 per cent and 7.63 per cent in 1911. With a 7 per cent rate, construction costs were paid for by the rents within 14 years (though buildings of this type are still standing today). It must also be said that during the period 1869–1911, the average interest rate stayed around 3 per cent, with a slight increase up to 4 per cent at the end of the century. By comparison the blocks of rented apartments were a profitable investment, even after deducting the various expenses, which came to less than 3 per cent. This explains why, from 1880 onwards, property companies began to appear and to expand.

An administrator of a property company providing popular accommodation in France considered that there had to be a 7.96 per cent gross yield so that he could service loans (3.5 per cent), pay management expenses (2.5 per cent), keep 1.06 per cent in reserves to cover repairs and assign the remaining 0.90 per cent to amortisation. During this period (1869-1911), most property companies were set up with no capital and so had to repay loans representing the full cost of acquisition out of their rents, even managing to pay small dividends. The Martin property company in Lyon, which was created with subscribed capital, was distributing dividends of 4 per cent to 5 per cent only a short time after it had acquired property. It is therefore certain that during the period 1869–1911 the property yield was above the interest rate. Most bankruptcies occurred because of unexpected increase in interest rates, together with maintenance costs that were higher than forecast, and an increase of vacancies in badly-located buildings, mainly in low-income districts. We can thus see that the rental yield cannot be defined primarily by derivation from the interest rate and that the latter is neither generator nor determinant of the provision of rented accommodation.

Abstracting from diverse rental forms and financial tricks, the property owner's cumulative earnings can be defined as follows (by year n):

$$Gr = \sum_{t=0}^{t=n} L - \sum_{t=0}^{t=n} E - \sum_{t=0}^{t=n} T - C$$

where

Gr = property owner's earnings
L = rents
E = maintenance costs
T = taxes and rates
C = initial investment: construction costs
t = time in number of years
n = number of years since construction

For buildings that have lasted for more than a hundred years, the mass of rent income is considerable, because the various yields quoted in the text above are yields on investment, not calculated on historic construction costs but on the building purchase price in which there is already a portion of capitalised rent.

Rental Yield and Rate of Profit

If we accept the conclusions of the foregoing analysis, it appears that rents repaid construction costs within about ten years (since maintenance costs were very low during this period and tax on housing was not very high). How then can expensive housing and high building rents be accounted for?

The notion of monopoly rent is excluded by definition; so is the effect of excess demand which could only have been intermittent. Yet there were unoccupied houses in Paris at the end of the century. Similarly, situation was not an important factor since most buildings were erected on the outskirts of existing towns, except those erected in the restructuring of Paris and Lyon. Urban infrastructures were confined to the town hall, the 'prefecture' and those three items dear to the bourgeois property owner; the café, the theatre and the opera. The cash flow of urban agglomeration could not be monopolised by one landlord and we must concentrate on the general attraction cast by a city wherein a community of property owners started to expand alongside industrial production.

We must assume that those investors who invested their money in property acted in accordance with the dominant capitalistic logic and that they were intending to derive a rate of return matching the rate of profit in industry. According to Markovitch's studies (1967), the rate of return on 'ownership and enterprise' in all industrial fields rose to 13.14 per cent in 1840-5 and 15.38 per cent in 1860-5. On this basis, we may suppose an average rate of profit on capital of 10 per cent. If this rate of profit is applied to the rent calculation, the rent figure will be 8F per 100 francs investment if the rate of profit is 8 per cent, 10F if the rate of profit is 10 per cent, etc. On this basis we can measure the time it takes for initial investment to be repaid by rent, assuming that the property owner is the sole investor of capital and that he only pays himself in the form of rents, which was precisely the case in the second half of the nineteenth century. The calculation is given in Table 8.2. We may call this rent, calculated according to the average rate of return and in constant prices, the equilibrium rent under a capitalist system.

Moreover, it should be pointed out that a rent calculated according to this method does not permit capital to accumulate at the same rate

Table 8.2: Number of Years' Rent to Repay Initial Investment at Various Rates of Return

Initial investment	Rate of return	Rent	Period over which capital is refunded
1	2	3 = 1 x 2	4 = 3/1
100 francs	0.08 (8%)	8 francs	12.5 years
100 francs	0.10 (10%)	10 francs	10 years
100 francs	0.12 (12%)	12 francs	9.3 years
100 francs	0.15 (15%)	15 francs	6.66 years
100 francs	0.17 (17%)	17 francs	5.8 years

as in industry. In fact, in a one-year turnover cycle, capital increase would be to 110 in the first year (100 X (1 + 10%)), to 121 in the second year (110 X (1 + 10%)), etc. In other words, even when based on average rates of profit, the rent yield is merely a perpetual annuity. This is the analysis which is generally used to examine such income and which justified various contemporary methods of calculation. In summary, the real rent yield will always be less than the average rate of profit. That is, in this period:

$$\pi > r > i$$

where π = rate of profit
r = rental yield
i = interest rate

The basic variable, rent, now having been defined, it is possible to calculate a real annual rent yield by dividing the sum of property incomes R by the number of years of life of the building, i.e. $r = R/n$. It is also possible to calculate a pseudo-rate of realisation j, by dividing the equilibrium rent L_o by the sum of the rent incomes R, i.e. $j = L_o/R$. By varying the life of the building, this method gives the results shown in Table 8.3.

We can see that the real rent rate effectively increases with the life of the building.

Using the same data, the effects of a variation of profit rates can be demonstrated. Consider for example a building with a 100 year life span, as given in Table 8.4. The real rent yield is also very sensitive to profit rate variations whereas the realisation rate is not.

These Tables 8.3 and 8.4 clearly demonstrate the two determinant

Table 8.3: Capitalist Building Rent and Rent Yield when the Life of the Building Varies

Case	n	C	π	L	n.L.	A	T	E	Σ ex-penses	R	r	j	d
1	25	100	10%	10	250	100	2.5	30	132.5	117.5	4.70%	8.51%	10
2	50	100	10%	10	500	100	5	80	185	315	6.30%	3.17%	10
3	100	100	10%	10	1 000	100	10	180	290	710	7.10%	1.41%	10
4	200	100	10%	10	2 000	100	20	380	500	1 500	7.50%	0.67%	10

Table 8.4: Real Capitalist Rent when Profit Rate Varies

Case	π	Co	Lo	n Lo	A	T	E	Σ ex-penses	R	r = R/n	j = Lo/R	d = Co/Lc yrs
1	5%	100	5	500	100	5	80	185	315	3.15%	1.58%	20
2	10%	100	10	1000	100	10	180	290	710	7.10%	1.40%	10
3	15%	100	15	1500	100	15	280	395	1 105	11.05%	1.36%	6.7
4	20%	100	20	2000	100	20	380	500	1 500	15.00%	1.33%	5

Key to Tables 8.3 and 8.4: n = life of the building; C = cost of construction; π = average profit rate; L = rents = C.π; A = amortisation; T = taxes and duties = 1% n.L; E = maintenance costs = 20% n.L. after refund of C by the rents; R = total income = sum of rents less various expenses; r = real annual rent yield = here, r/n; j = realisation rate = L/R; d = number of years needed to redeem the initial investment C by the rents.

variables: profit rate and the life span of the building (Gravejat 1981).

Where does the excess labour supplying this rent come from? If we go back to Marx's analysis, it is clear that this income stems mainly from the internal activity of the sector and that it represents only one form of differential rent II. The excess labour comes therefore from the wage earners of this sector and is clearly expressed in the rent. It is as if the growth of the urban population made it necessary to build new buildings, thus placing marginal accommodation on the market whose cost determined the average price of the product. As the production costs of the old buildings had been paid off a long time ago, the incomes they brought in must consequently be defined as rents. In this case housing can no longer be classed as goods, but primarily as property.

The Price of Building Land

Following this logic, the price of the land can be worked out by capitalising the future incomes after the building has been paid for by the rents. It can also be worked out just by capitalising the price of the building at that date. Financing costs must be added in the case of a landowner, because the fact of selling his ground means that he has not got the necessary money to finance construction. Depending on the methods applied, the price of building land will vary between 19.27 per cent and 29.14 per cent of the total price of the building without counting the promoter's interest payments. We arrive at a market price for building land equal to about 20 per cent of the housing price. This estimation seems to be quite acceptable.

Application 1: The Rented Building

This analysis provides an explanation for certain peculiarities of the rented building. Total income depends first on the level of the rent L_o, on the life of the structure n, on maintenance costs E and on various taxes T. Thus:

$$R_{max} = f(L_o, n, E, T)$$

The landlord cannot directly influence either the profit rate or the tax level. But he can make the construction denser so as to increase the equilibrium rent, he can improve the quality of the building; he can also try to obtain a higher profit by improving the appearance (e.g. decoration of luxury blocks of flats). He can also claim higher profits if the environment is attractive. Thus:

$$L_o = f(K_q, K_l, sP, sE)$$

where L_o = rent
K_q = invested capital
K_l = quality of the product
sP = higher profit where the quality is above average
sE = higher profit where environment is above average.

It is clear that it was no error of judgement to erect luxury buildings, assuming that the known demand is strong. Likewise, the landlord will try to ensure his building is well situated and derive an income from all the advantages of the city. This source of income was of secondary importance in the nineteenth century.

The rent yield also rises with the life span of the construction and the minimisation of maintenance costs. The landlord will thus select with care the materials, construction techniques, the potential of the space available and the situation. Thus:

$$n_{max} = f(M, T, f/p, loc)$$

where n_{max} = maximum life span
M = materials
T = construction techniques
f/p = functional potentials
loc = situation

The landlord will select dependable materials like stone, and resort to the competence of skilled craftsmen. It is noteworthy too that most architects drew their inspiration from classic or gothic styles, which did not go out of fashion.

Application 2: The Evolution of Construction

This analysis of the rent system in the production of buildings explains the evolution of construction in France. The system was basically unsuited to providing inexpensive accommodation, which explains the creation of an authority for a new production system at the end of the nineteenth century, called 'Habitations Bon Marché'. Subsequently, between 1914 and 1948, rent control completely wrecked this system and production was disorganised after World War II. The French government undertook to re-launch the sector, mainly through a policy of providing cheap rented accommodation. From the sixties onwards, property developers and banks were able to enter the market, resulting in an era of luxury buildings from 1965–73. Thereafter, the building industry was able to work according to its own commercial and technical needs, and individual houses, built with factory-made components, enabled a good quality, accessible industrial product to be offered on the market. Simultaneously, the average house price was based on the price of new houses and on this basis most old buildings could be rehabilitated so as to join the open rental market. It is thus, from this analysis of the production of rented accommodation, that we can understand the main choices of the government after World War II. It is also obvious that today there is no place for the private landlord. He can still act as a manager but he no longer finances construction of housing.

Conclusions

Starting from a study of the price of building sites, it appears necessary to re-examine the various recorded rent systems historically in order to understand not only the rent phenomenon in general but more especially the very specific character of capitalist urban rent. Such an analysis provides an explanation of the economic mechanisms and of the origins of the surplus labour which is the source both of building rents and of land rents.

References

Aboucaya, C. (1963) *Annales de l'Université de Lyon,* 3e Série, Faculté 24, Sirey, Paris

Carcopino, J. (1939) *La vie quotidienne à Rome à l'apogée de l'empire*, Hachette, Paris

Castells, M. (1972) *La question urbaine*, Maspero, Paris

Daumard, M. (1965) *Maisons de Paris et propriétaires parisiens au XIXe siècle,* Cujas, Paris

Derycke, P.-H. (1979) *Economie et planification urbaine*, Presses Universitaire de France, Paris

Garden, M. (1975) *Lyon et les lyonnais au XVIIIe siècle*, Flammarion, Paris

Gascon, R. (1971) *Grand Commerce et vie urbaine au XVIe siècle*, Mouton, Paris and The Hague

Goux, J.-F. (1978) *Eléments d'économie immobilière*, P.U.L. Economica, Paris

Granelle, J.-J. (1970) *Espace urbain et prix du sol*, Sirey, Paris

Gravejat, A. (1980) *La rente, le profit et la ville*, Anthropos, Paris

—— (1981) *Le prix d'offre des terrains à bâtir et la rente immobilière*, Thèse d'Etat, Université Lyon II

Guigou, J.-L. (1982) *La rente foncière*, Editions Economica, Paris
Guyot, F. (n.d.) *Essai de l'économie urbaine*, Librairie de Droit et de Jurisprudence, Paris
Homo, L. (1951) *La Rome impériale et l'urbainisme dans l'antiquité*, Editions A. Michel, Paris
Lefebvre, H. (1968) *Le droit à la ville*, Anthropos, Paris
—— (1974) *La production de l'espace*, Anthropos, Paris
Léon, P. (n.d.) *Géographie de la fortune et structures sociales à Lyon au XIXe siècle*, Centre d'Histoire économique et sociale, Lyon
Lescure, M. (1977) *Les banques d'Etat et le marché immobilier en France a l'époque contemporaine (1870-1940)*, Thèse, Paris X, Nanterre, 428ff
Lipietz, A. (1978) *Le tribut foncier urbain*, Maspero, Paris
Markovitch, T.J. (1967) *Salaires et profits industriels en France sous la Monarchie de Juillet et le Second Empire*, Cahiers de l'I.S.E.A., 'Histoire Quantitative de l'économie francaise', 4. Presses Universitaires de France, Paris
Marmy, M.-J. and F. Quesnoy (1866) *Hygiène des grande villes: topographie et statistique médicales du département du Rhône et de la ville de Lyon*, Vingtrinier, Lyon
Marx, K. (1973) *Le capital, III*, book 3, Editions Sociales, Paris (first published 1883)
Masson, M. (1980) *La rente, le profit et la ville*, Anthropos, Paris
Mumford, L. (1964) *La cité à travers l'Histoire*, Seuil, Paris
Neveux, H. (1971) 'Recherches sur la construction et l'entretien des maisons à Cambrai, de la fin du XIVe siècle au début du XVIIIe siècle, in *La bâtiment: enquête d'histoire économique*, Mouton, Paris, 189–312
—— (1975) *Histoire de la France rurale*, Seuil, Paris, vol. 2
Ricardo, D. (1970) *Principes de l'économie politique et de l'impôt*, Calmann-Levy, Paris (first edition 1817)
Schnaffer, B. (1957) *Les rents au XVIe siècle: histoire d'un instrument de crédit*, Editions S.E.V.P.E.N., Paris
Topalov, C. (1972) *Le capital, le propiété foncière et l'Etat*, Centre de Sociologie Urbaine, Paris
—— (1973) *Capital et propriété foncière*, Centre de Sociologie Urbaine, Paris

9 MARXIAN CATEGORIES AND THE DETERMINATION OF LAND PRICES

Agostino Nardocci

The aim of this chapter is to analyse the categories used by Marx in his analysis of land rent, in order to weigh up their importance within a particular, and to a certain extent narrow, range of problems: the determination of land values.

Marx on Land Rent

In a certain sense this objective is rendered easier by the fact that it corresponds to what Marx proposed when studying the economic value, the exploitation of land monopoly, within the capitalist method of production. One can start the analysis by asking two questions posed by Marx himself.

(1) Why is it that in agriculture the surplus value is divided into rent and profit, whereas in industry it is equal to profit?

(2) How can this happen if the portion of profit in agriculture is equal to that obtained in any other sphere of production?

An immediate reply can be given to the first question, referring to the monopolistic right of the landowner. In other words, the capitalist pays the landowner a sum of money to be able to use his field in the production process. The second question requires a more complex explanation in so far as it is necessary to refer to the differential rent of type I and II and to the absolute rent. Proceeding on these lines one runs into great difficulties because on these elements a single version accepted by all Marxists does not exist; on the contrary there exist contradictions which do not seem destined to vanish in a reasonably short space of time.

Even if in these controversies one can sometimes distinguish influences which have nothing to do with Marx's thought, there is no doubt that the origin of these interpretative difficulties must refer back to Marx himself. The causes are well-known: the sections which deal with land rent and its determination were laid out by Marx at different times, mostly in the form of notes (and sometimes even interrupted in their sequence) and – the final point but certainly not less important – they were never re-elaborated in a definitive form. This means that one has to try and follow the development of Marx's thought and to single out from this evolution the possible definitive outcomes, also having in the background the sources available to Marx.

In this light there are two authors of primary importance: Smith and Ricardo. In the case of the first, one must underline the contraposition affecting his analysis of land rent which on occasions is seen as a residual, on others as a cost. In the case of the second, it is stressed that rent is seen as a residual. Ricardo's contribution is so authoritative, in all its simplicity, that his influence is not only felt in Marx, but is still felt today in the economists of the neoclassical school.[1]

Let us now consider the method used by Marx to calculate the various types of rent. In this case we are presented not with one situation but with three different indications. In the *Theories of Surplus Value* we have the following:

$$
\begin{aligned}
RA &= VI - PP \\
RD &= PM - VI \\
RC &= RA + RD \\
RC &= PM - PP
\end{aligned}
\qquad [A]
$$

Where Marx calls RA absolute rent, RD differential rent, RC total rent, VI the value added in the single stages of production, PP the corresponding price of production and PM the market price of production (Marx 1969: ch. XII).[2]

Later in the *Theories of Surplus Value* in the summary tables and also in some indications advanced by Marx in the text, in certain situations for which the relationships of [A] would be valid, we have:

$$
\begin{aligned}
RA &= PM - PP \\
RD &= \text{zero} \\
RC &= RA + RD \\
RC &= PM - PP
\end{aligned}
\qquad [B]
$$

Obviously in this case the total rent coincides with the absolute.[3]

Finally the indications at the beginning of Chapter 45 of *Capital* (Marx 1972: ch. 24) which discusses 'Absolute rent', can only be generalised in the following way:

$$rd = p_a - p_i$$
$$rnd = p - p_a$$
$$rc = rd + rnd$$
$$rc = p - p_i \qquad [C]$$

where p_a represents the unit price of production obtained on the least productive field, p_i the unit price of production obtained on the ith field, p the price on the market and finally rd, rnd and rc represent examples of the differential rent, non-differential rent and total rent.[4] In a situation where the organic composition of agricultural capital is below the social average then the rnd becomes the example of total rent, and we will refer to this case until it is shown to be otherwise.

We now express the relationships drawn up in [A], [B] and [C] homogeneously. [A] becomes:

$$RA = (v_i - p_i)\,Q_i \qquad \qquad p = \frac{V_a}{Q_a} = v_a$$
$$RD = (p - v_i)\,Q_i \qquad \text{where}$$
$$RC = (p - p_i)\,Q_i \qquad \qquad p \geqslant v_i \qquad [a]$$

where v_i represents the value of the unit of production obtained on the ith field, Q_i the total amount of its production, p_i the unit price of production, and p the unit price on the market.

Following Marx's indications, the difference in productivity accruing to different plots of land can be set out in ascending order using the letters of the alphabet.

[B] becomes:

$$RA = (p - p_i)\,Q_i \qquad \qquad i < k$$
$$RD = \text{zero} \qquad \text{where } p = v_k \text{ for}$$
$$RC = (p - p_i)\,Q_i \qquad \qquad k = \text{index of leading field} \qquad [b]$$

while [C] becomes equal to:

$$RND = (p - p_a)\,Q_i$$
$$RD = (p_a - p_i)\,Q_i$$
$$RC = (p - p_i)\,Q_i \qquad \text{where} \qquad pQ_a = p_aQ_a + rQ_a$$
$$p = p_a + r \qquad [c]$$

One immediately realises that only RC can be established following the same method, whilst RA, RD and p changes as one considers [a], [b] and [c].

The distinctions and variations are not merely formal, but substantial, with important consequences for the analysis of land rent. It is worth considering the equations more carefully in order to highlight the more important results they give.

In [a] the determination of RA is shown to be independent of the differences in productivity of agricultural land and is instead a consequence of the discrepancy between the value and the price of production on each plot of land. Marx is therefore correct in considering RA as being completely different from RD, without there being any connection between the two types of rent. Even admitting that the market price corresponds to the value obtained on the leading field, Marx's reasoning only holds true if the market value is superior to the production value. When this situation does not arise there are only two possible alternatives. The first consists in maintaining the determination of RA unchanged. In this way it remains a function of the total amount of capital invested and its organic composition; nevertheless RD will present negative values, as Marx himself states.[5] The second alternative consists in assuming RD non-existent and to attribute all the rent created to RA. In this way one returns to the situation indicated in [b]. It should further be underlined that under these circumstances RA cannot be equal to each given amount of capital invested in the different plots of land, rather it becomes a function of the productivity of the land. This depends on the fact that reference to the market price, necessary for its determination, is connected indirectly to the quantity of the product obtained from the leading field and the fields under examination.[6]

It seems logical to presume that the unacceptable results obtained with this procedure (negative differential rents or absolute rents which are shown to be differentiated in the case of a zero differential rent) caused Marx to adopt [c].

It should nevertheless be underlined that in this case RA not only remains dependent on land productivity, identified in situation [b] through the connection with p, but accentuates it through its links with p_a, or rather with the price of production of the leading field.

Later Interpretations

Let us now consider the interpretations advanced by the followers and critics of Marx concerning these points. In them, as is easily noticeable, the influence of Ricardian analysis prevails and continues to dominate.

Differential Rent

In the traditional interpretations the specifications of the differential rent using the relationship indicated in [c] is only valid if referred to rent of type I, whilst it needs to be modified in the case of the differential rent of type II, in order to take into consideration the varying capital invested in the same plot of land. Proceeding in this way – although from a formal point of view it does not create any difficulties – has nevertheless substantial results on the rent itself, as Ball (1977) has demonstrated comparatively recently, in so far as the analysis is condemned to remain static without any chance of becoming dynamic.

In fact, if one maintains the Ricardian method, three insoluble problems arise.

(1) How can one specify a certain sum of money which becomes the unit of measurement of the subsequent investments?

(2) Why does that sum of money become the measurement of the investments made on any plot of land and not of those made successively on the same piece of land?[7]

(3) Why does one consider the investment made in industry as a whole, while in agricultural one has to divide it into segments?

To tackle these problems unresolved and unresolvable in Ricardo, was nothing more than the starting point which marginalist thought developed, as Emmanuel (1969) correctly points out.

It seems extremely important to me that recently, on this specific point, there has been a re-evaluation by the followers of Ricardo, with a complete volte-face of his method. To confirm this point one merely has to compare the position of Sraffa in 1925 and 1960.[8] As already stated the question is of importance, because if one follows Ricardo's indications one cannot help having a distorted picture of the productive process, and the price of the agricultural product will appear, for each given total amount of capital used in the agricultural sphere and each given level of productivity achieved, as a measure of scarcity – whether relative or absolute – of the land. At this point the land, in fact, is considered a factor of production, in the same way as labour. On the other hand if one follows the indications of [c] one arrives at completely the opposite conclusion. First of all capitalist ownership creates a discrepancy between actual production and potential agricultural production, through a constant block on the flow of capital to that specific sphere of production. In the second place it is precisely this block which prevents an increase in the productivity of agricultural labour.

So we may conclude that the existence of private property, in the historical tradition of the capitalist method of production, causes a less intensive use of social labour per hectare, compared to a society of united producers, and in that way it provides a relative 'scarcity' of land, or rather it cristalyses the differences in labour productivity applied in the sphere, creating and maintaining stable rent, which would otherwise tend to disappear. It seems significant that these very results can be obtained on the basis of neo-Ricardian and neo-classical paradigms.[9]

Absolute Rent

On this point criticism has constantly set its face against the very existence of absolute rent, on the basis of Chapter II of *The Principles* by Ricardo (1951). It is well known that it was necessary for Ricardo to eliminate land rent from at least one plot of land, to be able to determine the portions distributed between capital, labour and land. What is less known and what his followers disagreed on constantly is whether his procedure is correct. In fact his reasoning is based on a very restricting hypothesis, even unreal, but from a theoretical point of view plausible, granted the aims Ricardo had set himself. In a newly colonised country which he initially considered, there can be no rent, in so far as there is 'an abundant quantity [of land] *not yet appropriated*, and, therefore, at the disposal of whosoever might choose to cultivate it'.[10]

According to Ricardo the appropriation of land and its scarcity are closely linked, and yet what has to be underlined is the fact that in his exposition land rent is established in the same context as the formation of landed property. One can see that the very reference to a newly colonised country is only an instrument to eradicate the consequences of private ownership in a capitalist method of production. Followers, one knows, are not always able to overcome limitations, and what for Ricardo was a simple hypothesis to achieve certain objectives, became for them an absolute truth. And yet effective reality continues to insinuate doubts.[11] On the other hand the sharpest students, despite being in the continual midst of smoke screens, have abandoned the 'ipse dixit' as can be seen, for example, in the cases of Bortkiewicz (1910–11) and Meldolesi (1971).

Briefly, to synthesise all their arguments, especially those of

Bortkiewicz, a critical analysis is made of the relationships revealed in [a] and [b] and, paradoxically, they practically re-propose the findings of [c] as correct.[12]

And yet the conclusion is that non-differential rent, or rather absolute rent, does not exist. This procedure can only seem extremely incoherent in so far as there is a clear contradiction between affirming that private ownership of land can change the effective order of investments (or rather can momentarily raise the price of the product) and at the same state that this is due to the theory of differential rent and, furthermore, that it is of no great importance. The fact that the determination of the rent demanded for the last field cultivated considers the difference in productivity of the single fields has nothing to do with the generating cause of that rent. In reality, on the basis of declarations by both Bortkiewicz and Meldolesi, one has to deduce that dominant social relationships interact with the economic system and prevent the latter's full development. In these actual circumstances there is a discrepancy between how much the society can obtain, on the basis of the technology and quantity of capital used, and how much can be obtained in the same situation when private property does not exist (Nardocci 1982a). Whether the actual amount of rent demanded for the newly cultivated field is big or small is of no importance: the problem is theoretical and is dealt with as such.[13] Essentially one can affirm that these criticisms of Marxist analysis tend to strengthen the version identified in the relationships contained in [c]. It is, in conclusion, worth restating this and clarifying some further implications and the field in which this version can be applied.

Conclusions

Let us first deal with differential rent, and then pass on to absolute rent. With regard to the first, one can say that the most important implications were brought to light by Ball. Using the procedure based on [C], one discovers that the existence of landownership prevents a completely free movement of capital in agricultural production, and, therefore, the creation of the artificial monopoly.

Next to this fundamental result I must underline two others. The first is connected with the behaviour of the economic subjects and the logical steps they follow. In other words, Marx's analysis places the capitalists who are excluded from landownership in the primary position. Next to these can be located, in actual reality, elements which form in themselves the two types of capital and property. As I see it, the behaviour of the capitalist owners, without modifying the previous results differs from them in some ways. In particular I would state that if among these subjects the logic of capital prevails, the flow of social labour on the plots of land used will be greater, while if, on the contrary, the logic of property prevails, the use of land will be less intensive (Nardocci 1982a; 1982b).

The second element which I would like to bring up consists of a definition which arises directly from a statement made by Marx when he introduced the famous example of the energy obtained by using a waterfall, or coal.

If one studies carefully the numerical data used by Marx, they do seem to contradict the relationships expressed in [C]. In fact the total price of production of the commodities obtained with the use of the waterfall is equal to 103.5, while the price of production dictated by the market is equal to 115; with the result that the difference is equal to 11.5. And yet the rent actually paid to gain access to the waterfall is equal to 10, in other words 15 per cent less than the difference which exists between the prices of production. Now the supposed contradiction disappears as soon as one studies the examination of the situation considered by Marx. In reality if the economic conditions connected to the use of hydraulic energy have to be related back to those used by coal, one sees that the rent has to be paid at the beginning of the productive process and that the difference between the prices of production must be discounted on the basis of the average rate of the existing profit. Therefore, one can state that the RD will be the same as the difference between $p_a Q_i$ only when the rents are paid at the end of the entire process, while such a difference must be discounted on the basis of the average profits when it has to be paid out before the capitalists can even touch the agricultural land (Nardocci 1982c).

Examining now absolute rent or non-differential rent, two facts have to be stressed. However hard one tries, there is no comparison between analysis and reality, when the first is anchored in the hypothesis that there is at least one field used gratis. The norm, whether we like it or not, is represented by the owner who concedes the use of his or her plot of land, or by a capitalist who buys a plot of land to use in the productive process. In both cases cultivation is prevented from starting as soon as the price of the product permits the normal exploitation of the capital used on the new field. The second is the fact that, if the analysis is carried out with reference to the conditions posed by Marx, there is no chance of finding a possible mediation between the relationships indicated by [a], [b] and [c]. In reality, if one assumes a full development of the capitalist method of production, the complete separation of the landowner ('who can spend his whole life in Constantinople, whilst his land is in Scotland') and the industrial capitalist, there is not, and never can be, any possibility of referring to the individual values of each field. In fact the determination of the amount of r – that is, the surplus over the production price obtained on the field which regulates the market price – is linked, as Marx himself affirms, to the investment of capital which is possible on the fields already cultivated and, one might add, to the plots of land not yet in production, which have an even lower productivity. In this case one must deduce that the prices of production obtained and obtainable on the different plots of land become crucial in finding the amount of r, while no reference can be made to single values.

In the end it is worth considering briefly one last question: Is it possible to use the analysis of agricultural rent for urban land? It seems to me that the answer can hardly be completely affirmative. In the agricultural case one has a product which is indistinct and indistinguishable in the eyes of the consumer. On the other hand, in the urban case, each product unit assumes a specific character, on the basis of the characteristics given by the 'quality' of the product and the position – that is, the relationship it has with the rest of the urban unit.

Besides this there is another characteristic which must be stressed. In the agricultural case the intensity of capital investment does not need any social sanction, beyond the one given by the realisation of the production, and is therefore left to the will of the capitalist. On the contrary, in the urban case, it is rigidly sanctioned by public powers which decide the maximum density of capital per hectare. Therefore public power becomes a fundamental subject in understanding the differences which arise between the various plots of land which constitute the town.

Finally I do not believe that the category of absolute rent can play the role which much current analysis of urban rent has given it. It is probably more useful in order to explain the difference between the prices of agricultural land and urban land, to use the Marxist distinction between *terre-capital* and *terre-matière* rather than to refer to values of prices of production. So the difference in prices of the areas between town and country could arise from the concentration of fixed social capital which characterises the physical aspect of the first as opposed to the sparsity of the second. To proceed in this way makes the analysis much more complex in so far as the study cannot be limited only to the productive sphere, but needs to consider more carefully the effects which all the relationships between the various social classes induce in the capitalist method of production in its most mature phase.

Notes

1. One must only consider that these economists analyse the rent only as a residual; in von Thünen's model, and in his derivations too, the field which lies at the longest distance from the market does not provide rent.
2. Marx's study of the coal mines can also be applied to agriculture; Marx himself allows us to put quarters of corn in the place of tons of coal (Marx 1969: 2, 570).
3. See particularly the tables C, D and E (Marx 1969: 2, 574).
4. Marx's statement on differential rent and absolute rent makes sense only if we refer to rent rates not to total rents.
5. Cf. Marx 1969, 2, 589 and the tables C, D, E (ibid: 2, 571).
6. This situation is pointed out in the summary table D (Marx 1969: 2, 574) where fields I and II present RD which are zero and RA which are lower and distinguished from those which arise in the other fields. The differentiation of RA is also shown in tables C and E.
7. It must be also considered that Marx does not always refer to the total sum. See case B (ibid: 2, 568).
8. In the first paper Sraffa, while studying how capitalists act when operating the supposed principle of 'diminishing returns', supposes that the first investment in a field can be followed by others of equal amount. In the second work, on the contrary, when arranging the equations of production in the general system, he deals with labour and capital used on each plot of land *as a whole*.
9. Among neo-Ricardian scholars; see Montani (1972). In this remarkable paper the author argues that, given the corn price, the more productive technique will be used until land rent increases to a peak. It follows that the techniques which would increase the crop, reducing in the meanwhile (or even nullifying) the rent, cannot be used. Among neoclassical approaches see Nardocci (1982c).
10. My emphasis.
11. See for example, among neoclassical scholars, Needham (1981). Among neo-Ricardian scholars see Renata Targetti Lenti (1977). Needham states that landowners could use land for their pleasure, instead of offering it free; Targetti Lenti says that it is possible to assert that rent must be included in price determination.
12. It must be noted also that Bortkiewicz, who charges Marx with appropriating some of J.S. Mill's positions without quoting him, does the same to Marx himself. For example, he argues that in order to put under cultivation a new field landowners' requests have not 'to be excessive', but Marx had said it too when he wrote that 'a little production price increase will suffice to take to the market the worst new plot of land'. The first quotation is from Ladislaus von Bortkiewicz (1910–11); the second from Karl Marx (1972: ch. 45) (personal translation). Furthermore how is it possible that Bortkiewicz, who is a remarkable expert on Marx's theory, completely neglects the first part of the analysis of chapter 45 to devote himself only to the second part which refers to the arguments explained in *Theories of Surplus Value*?
13. As we have already seen Marx had also explained that the presence of this type of rent causes a small increase of production price. Nevertheless an increase

of unit price, though very small, involves a considerable sum when compared to total production. It surprises me that some phenomena are neglected because of their small size if in particular, one considers that 'modern economics' are largely based upon the concept of the derivative and, thus, of infinitesimal increase.

References

Ball, M. (1977) 'Differential Rent and the Role of Landed Property', *International Journal of Urban and Regional Research, 1,* 380–403

Emmanuel, A. (1969) *L'échange inégal. Essai sur les antagonismes dans les rapports économiques internationaux*, Maspero, Paris

Marx, K. (1969) *Theories of Surplus Value, II*, Lawrence and Wishart, London

—— (1972) *Capital, III*, Lawrence and Wishart, London

Meldolesi, L. (1971) 'Il contributo di Bortkiewicz alla teoria del valore, della distribuzione e dell'origine del profitto' in L. Meldolesi (ed.), *La teoria economica di Marx*, VII-LXX, Einaudi, Torino

Montani, G. (1972) 'La teoria ricardiana della rendita', *L'industria*, n. 3–4, 221–43

Nardocci, A. (1982a) 'La rendita fondiaria agricola in Ricardo e Marx', *Note economiche*, n. 1, 51–85

—— (1982b) 'La rendita fondiaria agricola e modi di produzione', Daest, IUAV, Venice, mimeo

—— (1982c) 'Un approccio neoclassico alla determinazione dei prezzi della terra', communication III Conferenza Naazionale AISre, Venice

Needham, B. (1981) 'A Neo-Classical Supply-Based Approach to Land Prices', *Urban Studies, 18*, 91–104

Ricardo, D. (1951) 'On the Principles of Political Economy and Taxation' in P. Sraffa (ed.), *The Works and Correspondence of David Ricardo*, Cambridge University Press, Cambridge

Sraffa, P. (1925) 'Sulle relazioni tra costo e quantità prodotta', *Annali di Economica*, 2, 277–328

—— (1960) *Production of Commodities by Means of Commodities, Prelude to a Critique of Economic Theory*, Cambridge University Press, Cambridge

Targetti Lenti, R. (1977) 'Risorse naturali, rendite e distibuzione dei reddit', *Giornale degli economisti*, n. 7–8, 419–53

von Bortkiewicz, L. (1910–1911) 'Die Rodbertus'sche Grundrentetheorie und die Marx'sche Lehre von der absoluten Grundrente', *Archiv für Geschichte des Sozialismus, 1,* 1–40 and 391–434

PART FOUR:
POLITICAL IMPLICATIONS

10 THEORY OF URBAN RENT AND THE WORKING-CLASS MOVEMENT: THE CASE OF ITALY

Vincenzo Bentivegna

One of the peculiarities of Marxist analysis of urban rent is the method. In most Marxist analyses of rent the main problem is to know how capitalism works in urban space and what is the role of urban rent, because this knowledge is important to understanding the capitalist mode of production in the present phase of development. Urban rent is seen as a component of a wider process in which masses of people act and struggle and where contradictions are becoming more and more explosive. Therefore, Marxist analysis cannot aim simply at the knowledge of mechanisms; it has a central point the understanding of social relations as they appear in the city.

For this reason I suggest that the interconnections between urban rent theory and the political and social praxis of the workers' movement is an essential field of analysis.

In the Italian situation in the last twenty years, these interconnections have been so strong in our field[1] that, as subjects for analysis, they legitimate each other. This is not casual, because this reciprocal legitimacy found its base in the general strategy of the workers' movement on the questions of housing and land. The roots of this strategy lay in a redistribution hypothesis: on the assumption of continuing growth of wealth, left policy on housing and land tried to link the pursuit of economic efficiency, social progress and social equity without effectively contesting either private ownership of the means of production or the private control of the accumulation process in the building sector. In exchange, the working-class movement required the guarantee of state intervention in wealth and income redistribution, in order to assure better life quality in towns, the spreading of collective services, access for low-income people to the housing market, jobs in the building sector, and so on.[2] The theory of rent which dominated the Marxist culture of that period – and became the scientific background for class-movement praxis – is coherent with this political strategy.

One of the most important questions in the Marxist debate on urban rent is the way in which urban rent is integrated within the capitalist system. If it is considered as an anomaly, rent can be considered a strategy completely antagonistic to capital. This means that urban rent is an abnormal factor in the system, a transitory phenomenon which survives thanks to its ideological features and the power maintained by landowners within the social formation. If, on the contrary, rent is considered an organic feature of capitalism, this means that rent is a structural phenomenon, thus compatible with the system and an expression of its being in this phase of development. In view of the impact which the scientific formulation has on the choices of the working-class movement and its policy, this question is of more than theoretical importance. It is very important politically: the primacy of one concept or the other leads to completely different praxis.

In the period I am referring to, the first was the dominant interpretation among Italian Marxists. Starting from the assertion that land is an external but necessary condition for capital, the analysis focuses on the peculiar features of its value, for use and exchange, the institutional privileges which permit landowners to appropriate rent, and the class characteristics of this property are isolated on the basis of abstract schemes of social categories. Landed property is thus regarded as an autonomous class, the heritage of a feudal mode of production preserved within the capitalist mode of production. Its activity expresses itself only at a distributive level and its aim is to maximise the opportunities for, and the levels of, rent against profit and social advantages, obtained by means of interventions in the city by capital or by the state. Therefore this class is identified as autonomous within the social formation, as an antagonist both of the interests of capital and those of 'the community'. Since the process of transfer of surplus profits into rent is obtained only via the right of property, this social class operates passively and rent has no function within the process of accumulation, except some interference with the building cycle. Therefore the behaviour of the agents in the market becomes important, since the contradiction is confined to the distribution of surpluses between landed property and capitalists.

The share of rent, which the landowners appropriate, is traced back to the action of capital or of the state. But the discussion on capital is limited by the shares of surplus profits immediately transferable into rent and therefore to the building process. The analysis of capital is based on a competitive concept. In fact, if the process of production is considered outside the rent model, the dynamics of competitive capital flows lead towards a harmonisation of rates of profit. This is obstructed by landed property which favours a difference in these rates, between the sectors (absolute rent) or balances the rates of profit within the sectors (differential rent), but in an 'abnormal' and negative way because it diverts excess profits from productive re-use. In the process of distribution, monopoly capital, in the form of financial capital, is interesting for its relations with landed property and with the building cycle, but is regarded as a factor capable of hindering or favouring the maximisation of rent, depending on the circumstances, without changing its quality.

The question of the state is more intricate. On the one hand, the state takes the responsibility for the main interest of the entire capitalist system (it safeguards the rights of property and mediates the conflicts between social classes for the possession of surplus profits); on the other hand, it reproduces the general conditions of production, regulates the use or urban land and supports a part of the cost of reproduction of the labour force, but the advantages of its intervention tend to flow into rent. In a static framework limited to redistribution of excess profits, if the state is seen just as a redistributive mechanism, its capacity to intervene in the production process is quite outside the model. Thus a reformist interpretation (state as an abstract producer of organised territory) and a radical interpretation (integration of the interests of landed property with those of the capitalist state on the basis of economic parasitism and of a system of conservative values) are not in contradiction. Generally speaking, the question of rent is identified as a problem of use of urban space and not as a consequence of the continuing transformation of urbanised space, so the urban crisis is traced back to landed property and to its capacity to collude with the state. Rent is therefore the source of contradictions in capitalist cities.

Thus urban rent is mainly considered as a question of subtraction of incomes and money capital surpluses without effects in the field of production (except for its limited impact within the building sector), an obstacle to the valorisation of capital and a fundamental cause of the contradictions in the city. Rent represents an abnormal and transitory phenomenon of the system, which could be eliminated by the rationalisation of the state and capital. It is an economic expression of the effects of the right of property on urban land, whose presence within the capitalist mode of production in this stage of development can be explained only in ideological terms and by the interlacing of interests between landed property and conservative forces.

This logic, based on the distributive approach, is characterised by the fact that dynamic variations, except those produced within the model, are not accepted. This excludes interferences which originate from the process of accumulation, because it is not necessary to refer to the dynamics of the categories and of the wider system to explain the formation of rent as a form of income. But to abandon every relation with the production of surplus value inevitably leads to concentrating attention not on the consequences of the changes in the process of valorisation of capital, but on those occurring in the process of redistribution of the surplus value. This undoubtedly leads to interesting considerations; but it diverts attention from the objective characteristics of the dynamics of the rate and mass of profit produced in monopolistic conditions, and the consequences for the transformation of surplus value into profit and profit into rent. So to limit the analysis of rent to the sphere of distribution implies an underestimation of the structural connections between rent and profit and between landed property and capitalists. It permits the construction of microeconomic and static schemes, mostly descriptive and limited to the analysis of the mechanisms.

This logic has consequences for the interpretation of the political meaning of urban rent. In fact, if we remain within the distributive perspective, the main point becomes the possession of the shares of excess profits, and the fundamental contradiction of rent develops (depending on your standpoint) within the bourgeoisie (between the capitalist class and the class of landowners and between the various fractions of landowners) or between landed property and the 'community'. The role of the working class thus fades into the background and it is concerned only to secure some of the available surplus. It aims at affirming its right to safeguard its reproduction and satisfy its needs. In fact in this case the way in which surplus value is appropriated does not imply the exploitation of the working class, since there is no direct relation with the process of production. The reduction of the role of the proletariat only to concern over levels of consumption omits the relationship between the struggle for rent, the struggle for the control of land use and houses and the fundamental conflict between classes over the production of surplus value. It thus favours a limited and substantially reformist vision of the class struggle. In this sense the control of the process of production and reproduction of capital in the city is reduced to the control of landed property which, as an antagonist to capital, can be isolated. It is thus possible to reach an alliance between capitalists and the working class for a better distribution of incomes or to entrust the same objective to the state, 'the rational institution'. It is evident that if urban rent is made dependent on landed property as a relic of the feudal mode of production sitting within the capitalist mode of production, the elimination of private landed property leads to the elimination of rent.

If we assume that this elimination can be brought forth by the state by nationalisation, absolute rent disappears as well as the effects of differential rent, because the surpluses which form the latter type of rent could be applied for productive uses. The consequences of this operation would be a more rational use of urban land (thus obtaining 'better' cities) and the development of the building sector which would be able to compete with the other productive sectors in terms of organic composition and so on.

In Italy, this interpretation of rent has had an important political meaning because it was an axiom in the urban policy of the working-class movement until the end of the seventies, as a component of a wider strategy on housing and urban questions.

From the economic and social points of view, emphasis is given to the cost of reproduction of labour forces and to the quality of urban development induced by landownership (which class is seen as absolutely conservative) and attention is paid to land and housing costs. According to the theory summarised above, this means paying attention mostly to the circulation process, both in land and housing markets (so land values become very important) and to land use (urban planning and expropriation being considered the best tools to control it).

It is important to underline that, substantially, this interpretation of urban rent leads to theoretical results which do not differ very much from those reached by reformism, even if the theoretical frame, approach and methods are different (Lombardini 1965). Reformists also show that negative effects of urban rent lie on the redistributive side (house prices structurally higher than marginal costs) and on the urban question (speculation, centrality of rent in the urban crisis, land monopoly as the allocative factor in land use, etc.). They, too, see the demiurge role of the state, able either to overthrow these negative effects (state as 'super parties') or to control the absolute conservatism of landowners (ethics of progress supported by the two productive classes). Therefore, the main political feature is that an alliance is possible between capitalists and the workers' movement to reduce the power of the landowner class and control the negative social and economic effects induced by rent on the redistributive side (more houses for all, more collective services, higher quality of life in towns, etc.).

The problem is that this alliance – by its very nature – cannot intervene in the behaviour of firms and financial capital.

Failure of the Alliance

The social and economic results achieved by this alliance are not to be underestimated in the Italian context. Italian towns have become more rational: townplanning is generalised all over the country; collective services are widely distributed; urban rent is under control in the public housing sector; speculation in the housing market is now much more sophisticated; landownership by an autonomous class is no longer a major problem in our cities. But the main goals of the workers' movement in the sector have not yet been reached: housing prices are still rising faster than the capacity of lower income groups to enter the market; rent is still the main factor in the allocation of urban land and, moreover, capital has now taken the place of the ancient forms of landownership, having particular interests in developing large schemes, tertiary sector buildings and tourism (Instituto Gramsci 1983). The problem is that the strategy was shortsighted and its rent theory inadequate.

This alliance was realised at state level and it was a part of the general tendency of that period both to remove obstacles to accumulation and productivity in the economic system and to redistribute wealth on a larger class base.

If we read this rent policy as a component of the general redistribution policy carried on by the workers' movement in housing we can see a model of state intervention which has many social democratic features (control of financial flows to support low incomes, diffusion of social services, incentives in the housing sector only on the demand side; agreement between productive forces against social privileges, etc.).

In fact, the emphasis on redistribution in the housing sector as a basis of the alliance implies:

(1) State participation in housing and landownership and state control in land use, but limited to operating either with the same rules as private capitalists or in such a way that structural mechanisms of the market will not be challenged or disturbed.

(2) Substantial state support to costs of reproduction of the labour force, especially for the production of goods and services which do not enter the direct relationship between capital and labour.

But that means that urban rent must be put under control only because it is: (1) a component of the cost of houses and collective services; (2)

a location factor for houses and services.

In other words, the roots of the rent policy which came with this alliance implied accepting that the state unifies social conflicts through public spending and supports accumulation, acting against all archaic economic phenomena like urban rent. The theory is then coherent with the strategy and vice versa. But this means that the workers' movement gives up any possibility of discussing the production side of the building sector. The workers' movement, in fact, does not have its own model of the town in which to place its own theory of rent.

Now the political conception which was at the base of the theory of rent as an 'anomaly' of the system is itself in crisis. On the one hand, with the crisis of the Italian form of welfare state, the cultural, social and political background of that alliance between workers and capitalists against rent is no more. On the other hand, the old theory of rent is no longer able to explain the mechanism of rent formation and extraction or the dynamics of rent in an urban world where rent has become a field of manoeuvre for financial capital.

A New Theorisation

The problem of urban rent has now to be tackled in a different way, starting from the analysis of the functions of agents in the dynamics of categories and of class struggle, thus considering rent as a structural feature of capitalism, at least in this phase of its development. The most important methodological transition has been to emphasise the dynamics of capital, or better the dynamics of surplus value and of profits and thus the valorisation of capital, by focusing the analysis on production instead of on distribution (see Ball 1977; Massey and Catalano 1978; Topalov 1981).

Following this discussion, the problem of landed property is no longer an ideological and institutional question, but it becomes the problem of the roles played in the capitalist social formation. The attention is therefore captured by the economic and social relations which comprehend landed property itself. But even if it is not a necessary condition for the existence of capitalism, it undoubtedly is an historical necessity which is part of the dynamics of the system, and can assume the most adaptive forms for the entire process of the development of capital. Rent, in its turn, can assume various roles and forms according to the type of property. In this case the contradictions belonging to the control of the material conditions constraining capital and control of the surpluses are different in form and in quantity according to the solutions adopted by capital, but they are not totally surmountable by the simple elimination of a particular system of urban landed property. In fact the effects of the possession of rent depend on the existing relations within landed property which, in their turn, are a function of the specific relation between rent and profit in the process of accumulation and, in the final analysis, they depend on the material conditions of capitalist production and reproduction. The specific contradictions between rent and profit – which originate from the complex ones of capitalist development – lead to the formation of new forms of property which also have contradictory effects, although different, on the process of accumulation. State property, in particular, introduces new contradictions although different from those of landed property.

Contemporary landed property in the city is characterised by its integration with and subordination to capital, in the sense that the evolution of the capitalist system has brought an evolution of urban landed property articulated in several fractions. On the one hand the part of landed property integrated and organic with financial capital which dominates the urban system; the majority of owners who, subordinated to financial capital, have very little control over urban development. This is not to say that this latter form of landownership is not profitable, but that the choices of such owners are strongly conditioned by the leading fraction of capital. Therefore the characteristics and functions of urban rent in particular cases depend not only on the mechanism which has brought their formation (absolute, differential or monopoly rent), but also on the position of each fraction of landowners in the social formation, in political, economic and ideological terms.

The characteristics of class autonomy tend to assume largely formal guise which is subordinated to compatibility with the process of valorisation of capital. The form in which the antagonism between landowners and capitalists now occurs depends on precise, historical eventualities and landownership tends to assume subordinate and controllable patterns, appropriate for the managing of rent under the control of capital by means of the progressive domination by capital of the entire economic and social formation.

Capital is the central category of the analysis, and urban rent is considered in terms of rates of surplus value and of profits, both in respect of its formation and of its function in the process of accumulation. Thus, purely competitive concepts of capital can no longer be proposed, and a realistic view must be accepted of the domination of the monopolistic fraction over the dynamics of the process of accumulation. Thus rent can be considered in terms of the dynamics of the categories and of the system (the tendential fall of the average rate of profit, the devalorised social capital, the role of the state in the dynamics of the system, production and dynamics of fixed urban capital, development and underdevelopment, etc.). The role of landed property in the formation of constant surpluses of the sector and of the firm follows the logic of capital, while its possession becomes more and more a problem of integration and subordination of urban rent to financial capital. This confirms that rent has its own structural effect which is more widespread than that obtained by limiting the analysis to the mere building cycle. In fact this set of relationships not only involves the financial circuit, but also directly influences the production of surplus value and the rate of accumulation.

Even in this case rent does not imply any controls on the process of production and represents an intervention in the distribution of surplus value which is created by the competition (between the fractions of capital and between single capitalists) for the control of the conditions of reproduction outside the process of production and for the possession of excess surplus value. Only the compatibility with the process of valorisation of capital changes (via the role of monopoly financial capital). From this point of view, rent is considered a contradiction of the system. The productive choices and the entire development of capital determine the conditions for the production of rent; rent is thus an internal event which depends on capital even if it is different from capital. Urban rent, in all forms, is an obstacle for the development of the productive forces used on land and, so being, it is contradictory to the accumulation of capital. This contradiction, however, is within the capitalist mode of production, because capital is capable of controlling rent itself; therefore the tension between landed property and capital does not necessarily imply the elimination of rent, but its harnessing to the valorisation of capital via a closer connection with the structures of financial capital. This does not eliminate the contradiction between rent and capital, but modifies its qualities and impact by placing the conflict which once existed between landowners and capitalists within the capitalist class.

An overturning of the traditional conception is in this way achieved. The affirmation of the role of rent in the process of valorisation of capital brings as a consequence that the types and forms of landed property – the pre-condition for the formation of rent – are mostly defined by capital according to the types and forms of rent compatible with the historically-given situations, and not vice versa. It is a process in which capital takes possession of urban rent and then adapts it to the forms necessary for the control of the conditions of production outside the cycle, the development of accumulation and formal, juridical institutions. This leads to a conflict between the old forms of property and the new ones. However, the levels of conflict and the transformation of the existing forms of landed property into new ones are connected with the particular conditions in which the process takes place. Rent is no longer only an irremediable, even if secondary, contradiction, but it is placed with its own specific function in the immanent contradiction of the whole capitalist system – between the development of the productive force of capital and the growing difficulty of using this productive force to enlarge reproduction.

The role of the capitalist state is now inside the model because it conditions the levels and the types of urban rent in the light of the entire process of reproduction of the social system. It is thus possible to recognise a general state policy. On the one hand, it tends to mediate between profit and rent and to reproduce the general conditions in terms of valorisation-devalorisation of capital. On the other hand, the state facilitates the drawing of rent by financial capital by imposing a capitalist logic on the use of urban land. In this way the state becomes a planner of the conditions of formation and creation of rent, but with the aim of supporting the entire process of accumulation of the system and to control its main contradictions. The production of general conditions and the control of urban land – and thus the conditions for rent – occur in terms of accumulation and not in terms of rent and rent is, in a long period of time, a mere by-product – even if particularly desired – of the capitalist process of production of urban space which occurs in terms of industries and services, concentration and differentiation, development and underdevelopment, slums and waste, etc. And thus we see the new relationship between capitalists, state and landed property for control of rent.

Within the urban system, determined and defined by the process of valorisation of capital, rent is capable of regulating the use of urban space via the price of the land and by influencing the location of capital and of the labour force; it is also capable of conditioning accumulation in the building sector by influencing the production of urbanised space. The relation between production of urbanised space and creation of rent is not necessarily complementary, although one derives from the other; it can also be contradictory and can produce conflicts, the results of which can only be interpreted by an analysis of reality without abstract or generalised preconceptions. In this sense the urban crisis is the consequence of the structural crisis in the process of capitalist development, with its specific historical, social, economic and political qualities. The effects of urban rent are a consequence, more or less important according to the situation. However, rent does not determine the urban crisis, although it is able to augment its effects.

A New Praxis

Rent is considered in terms of class relations. The role of the working class in the conflict for rent is no longer only connected with the satisfaction of its needs but is the expression of its capacity to effect the production and consumption of land as an element in the general conflict over production and distribution of surplus value. The alliance between capitalists and landowners, even if we accepted the latter as an autonomous class of the bourgeoisie, is based on the process of accumulation as a whole. In this way we can bury those extreme interpretations which, on the one hand, have brought us catastrophic visions of rent as an element of the vast problem of housing (the impossibility of a capitalist solution to the housing problem) and on the other, a reformist optimism which considers urban rent eliminable by means of state policy.

This interpretation of urban rent does not allow any further alliance between workers and capitalists against the landed interest because the problem is no longer to change within a redistributive model. The struggle is on the role of rent and of land in the process of accumulation: to limit and control the behaviour of firms and financial capital. The state cannot be seen as the focus of an alliance but becomes a focus of struggle. The central question is no more the control of public expenditure towards redistributive goals or of urban changes via master plan but the changing of the rules of the game and of social relations on the ground – in other words changing the whole accumulation process.

Thus rent, as a formal category, loses a lot of the political, social and economic importance which it has had during the last twenty years, in favour of the accumulation process in the building sector and the process of transformation of land use. The main questions become the character and the limits of building sector accumulation and the transformation and use of land as a fundamental resource (Bentivegna 1983).

Notes

1. On the roles and effects of urban rent on the economic system and the building sector see Ceri (1975), Insolera (1973) and Cervellati (1976).

2. On the political strategies of the working class during the sixties and seventies see Barcellona (1983).

References

Ball, M. (1977) 'Differential Rent and the Role of Landed Property' in *International Journal of Urban and Regional Research, 1,* 380–403

Barcellona, P. (1983) 'La crisi della politiche di programmazione' in *Critica Marxista, XX,* no. 4

Bentivegna, V. (1980) 'La questione della rendita nella teoria marxista contemporanea' in *Critica Marxista, XVII,* no. 4

—— (1983) *L'interpretazione della rendita urbana: l'esperienza italiana*, Instituto Gramsci, Bologna

Ceri, P. (ed.) (1975) *Casa, Città e struttura sociale*, Editoriale Reuniti, Rome

Cervellati, P.L. (1976) 'Rendita urbana e transformazione del territorio', in V. Castronovo (ed.) *L'Italia contemporanea*, Einaudi, Turin

Insolera, I. (1973) 'Urbanistica', in *Storia d'Italia*, Einaudi, Turin

Instituto Gramsci (1983) *La questione della rendita in Italia*, Bologna

Lombardini, S. (1965) 'Considerazioni sulla rendita edilizia' in F. Forte and S. Lombardini (eds.), *Saggi di economia*, Guiffré, Milan

Massey, D. and A. Catalano (1978) *Capital and Land: Landownership by Capital in Great Britain*, Arnold, London

Topalov, C. (1981) *Le profit, la rente et la ville*, Centre de Sociologie Urbaine, Paris

11 PLANNING AND THE LAND MARKET: PROBLEMS, PROSPECTS AND STRATEGY

Michael Edwards

Introduction

In the history of capitalist Britain the problems and contradictions associated with forms of private landownership have been of recurrent importance – in the material conditions of life, in the political sphere and in the theoretical focus of economists' work. The period since World War II has been no exception. The urbanisation component of the long boom greatly extended the housing stock (see Chapter 3) and expanded and reorganised many other features of the built environment. But we are left with severe 'inner city' problems, with a resurgent housing crisis, with extensive unused land and labour and with a construction sector still fraught with problems both for capital and for those working within it. We have seen three Labour governments legislate on landownership, with fierce political debate on each occasion. And intermittently through the period researchers and intellectuals have concerned themselves with these matters.

It is the contention of this chapter that the intellectual and political treatment of land questions has been substantially abortive. This is attributable in part to a failure to examine issues in an adequate theoretical framework. The whole of the present book substantiates this criticism and proposes an alternative. In part, too, the inadequacy of work on land has reflected a very narrow perspective: the failure to consider a wide enough context.

The first section of this chapter argues that the context for a political economy of land in the mid eighties is strongly conditioned by (i) the material shift in production and consumption relations from a period when some modest planning and political control existed to a situation now where financial criteria and commodity relations dominate the material conditions of life and have great power in the realm of ideas; (ii) changes in the spatial division of labour and a weakening of the power of organised labour, with strong consequent effects on the geography of class conflict; (iii) major change in the scale and nature of financial institutions, in the character of control over production and investment and consequently in the clarity with which class interests are perceived. The clarity of social relations is further obscured by the growing complexity of the overlap between 'public' and 'private' spheres of the economy. The second section highlights the mistaken formulations of the land questions themselves within this context and suggests the basis for a fresh approach. The chapter concludes by proposing the minimum ingredients which should comprise any future attempt to transform landownership.

The Economic and Political Context

The formation of a modern view of rent and landed property takes place in a specific context. This context, and recent economic and political changes, help us to specify the scope necessary for an analysis of rent and the politics of land questions.

During the long post-war boom the UK experienced a relatively stable regime of social democratic government including three periods of government by the Labour Party. Major expansion of the public sector took place, both through the growth of established public services – education, health, social services – and in nationalised industries. The growth of the public sector made a major impact both in drawing more people into the paid labour force and in the growing provision of housing and welfare services as the state became steadily more deeply involved in the reproduction of the labour force. While the war-time machinery of state economic planning was largely dismantled by the early 1950s, a number of important planning mechanisms have operated during the period, of which the land-use planning system created by the 1947 Town and County Planning Act is the most familiar. Alongside it central government controls and incentives aimed at regional redistribution of new manufacturing investment have had varying importance. Sometimes even the internal planning of central and local state expenditure and policy has been co-ordinated.

At the level of the economy at large governments, both Labour and Conservative, attempted various forms of consultative and indicative planning exercises with private capital and with the trades unions organised through the Trades Union Congress (Budd 1978; Glyn and Harrison 1980). The combined effect of all these planning measures meant that, in the late 1960s and in the 1970s, large parts of economic and social life in the UK were, at least nominally, subject to a measure of public control or influence. This is not to say that the operations of the state and its agencies ran counter to the interests (somehow defined) of capital, nor that there is anything inherently progressive in the way labour was controlled and organised within the state sector (Whitfield 1983: 77--9; Joint Trades Councils 1982). What it did mean, however, was that non-market decision criteria could rule – in some circumstances and spheres – and that local or national elected bodies could be pressed to act in ways not indicated by narrow financial criteria, that is, in pursuit of social need. Thus for example in land-use planning we had decisions which sought to take account of 'environmental' considerations and cases where organised campaigns (e.g. by tenants) overcame strongly-articulated pressures from capital (Palmer 1972). Public housing authorities had at least the potential scope for coherent social management of their large and diverse stocks which could be allocated, at least in part, on critieria of need.

Against this background the monetarist policy and ideology of governments since the late 1970s have wrought a major qualitative change as well as a simple shrinking of (non-military) state activity. The quantitative changes are familiar and well-documented by now (Glyn and Harrison 1980: 119, 140; Millwood 1981). The qualitative change has been the enlarging of the sphere in which goods and services are traded as commodities, in which users become individualised in their 'market' relations with producers and providers. The sale of council housing, the shift of housing association lettings towards market-dominated rents (Hodgkinson 1982) and the payment of housing subsidies to individuals instead of to the housing authorities are all components of this process in the housing field. In transport planning the old struggle between vested highway interests (the 'road lobby') and the advocates of public transport has largely halted, with private market criteria strongly to the fore in all important fields, save in a few metropolitan areas with embattled socialist administrations. Similar qualititative changes are affecting most spheres of socialised production (Direct Labour Collective 1978, 1980) and provision (Whitfield 1982: chapter 5) and are an important part of the context in which land questions are viewed in the next section. Market forces seem more inexorable, the 'feasibility' of projects for private capital a more tolerable critierion, when these abstractions are echoed by the changing material realities of daily life.

An equally important part of our context flows directly from the intensified exploitation of a vulnerable and weakened workforce. Structural economic change, the enlargement of the reserve army of labour and the downward pressure on real wages have begun to undermine the old spatial variations between regions (and perhaps between towns and countryside). In most occupations and areas the worker can believe in the threat from the unemployed applicant who would so readily replace him or her. This feature of the crisis, combined with anti-union legislation and the accelerated decline of old and highly-unionised sectors, has been a major setback for organised action and has eroded one of the dimensions on which British towns and regions have long differed. The rapid process of deindustrialisation and the growth of the service sector yielded an important change in the spatial scale at which social differences appear. Massey (1979) in particular has shown how the very 'coarse' division of labour inherited from the nineteenth century has largely been replaced by a more subtle and a 'finer' division of labour between city and suburb, between large and small towns and between metropolis and provinces. (See also Massey 1983 and Edwards and Bentivegna 1983.) The economic map of the country is very different from that of thirty years ago and it is this new map which capital confronts in its drive for accumulation.

The institutions through which accumulation is governed have become more complicated, and more politically problematical, in two important ways – through changes in the financial sector and through changes in the way public and private capitals relate to each other.

The relevant major change in the financial sector has been the growth of the insurance companies and pension funds. The long-term trend to the aggregate dominance of financial capital has been discussed and documented widely (see Coakley and Harris, 1983 especially chapter 8 and bibligraphy). A reinforcement of this tendency has flowed from the decision of the government to opt for a fully 'funded' system of occupational pensions. This is a system in which workers and their employers make compulsory payments into funds, and it is the profits and capital growth of these funds, invested through the capital market, which secure the retirement incomes of the same workers. As Murray (1983) has pointed out, this breaks our direct dependence as workers on the product of future labourers for our maintenance in retirement and replaces it with an indirect dependence via the continued long-run valorisation of assets. (The alternative would have been to stick to an explicitly state-operated scheme in which current taxes pay for current pensions.) This new mechanism now has two important consequences, one material and direct, the other ideological, political and indirect.

The direct consequence is that very large compulsory savings, as well as some voluntary savings, are channelled through the private insurance companies and the semi-private pension funds which run the schemes. Marxists disagree on the scope for radical transformation of these funds (see Murray 1980; Minns 1980), but there is no doubting their economic power as controllers of investment flows and of the production (and other) enterprises in which they invest. They are, of course, powerful influences on government, to whom they are also major lenders.

The indirect consequence of these institutions lies in the fact that, if my pension depends on the future value of the capital assets (and to some extent the property assets) in which my contributions are invested, then my interests as an individual worker – and as a member of the class of workers – no longer seem to be so antagonistic to the interests of capital: indeed my pension seems to depend on the long-run profitability of capital. It is a powerful device and it echoes the way widespread owner-occupation of housing seems to give the owners an interest in rising property prices.

In writing about pension funds just now I described them as 'semi-private' because they have some of the attributes of private corporations – in particular their beneficiaries are private individuals and the funds are run strictly in (what are believed to be) the interests of those individuals. But they also have some collective attributes, being run in many cases by boards of trustees with substantial worker participation and with close regulation by the state (Minns 1980). This rather indeterminate position between the 'private' and the 'public' (i.e. state) realms is increasingly important in the UK today for there is an increased blurring of the old familiar boundary.

A number of forces appear to have contributed to this blurring over the last decades. The provision of infrastructure and the development of urban areas, the execution of research, the training of the labour force have all taken on an increasingly social character: private capital unaided has been less and less able to carry out these operations profitably and the burdens have shifted to the state – central and local. Many other spheres of production have exhibited the same tendency – agriculture, steel, the nuclear power industry and much of transport are the obvious examples. This is an important facet of the growth of the state machine and of state expenditures – both direct spending (e.g. on training, on roads) and indirect (on subsidies to capital). The long trend to increasingly social production has been paralleled in the last few years by an increasingly private pattern of appropriation of the social surplus. In other words ways have been found to channel the surplus generated in state-supported production towards private beneficiaries – individual and corporate. The mechanisms which achieve this are diverse and include the admixture of state funds in essentially private corporations, the invitation of private investors into essentially state projects, the establishment of joint public-private bodies, the 'privatisation' of national and local government services by sale, by subcontracting or by shareholding – and many other devices. In these cases we frequently observe that old public and private assets (factories, docks, hotels) are effectively being devalorised at collective expense: their exchange value is being partially written off to the point at which they can be bought so cheaply that private investors can run them at attractive rates of return. This happens not least in the sphere of land and urban development, for example in the redevelopment of the London docklands.

All these factors are outlined in relation to the UK but all must find at least echoes elsewhere in view of the increasingly international character of the accumulation process. Recently this internationalisation has been both well described and analysed (for example by Harris (1983) and by Sutcliffe (1983)) and extensively theorised (for example by Harvey 1982) in relation to urban questions. For our purpose it is only necessary to stress two major distinctive features of the UK, both important for the land question: first the remarkably rapid deindustrialisation of the economy compared with other OECD countries and secondly the strongly-developed international flows of investment, particularly since exchange controls were finally removed in 1979 (Currie and Smith 1981). Activity in the UK is being rearranged anew as part of the international division of labour and the international investment market – at what appears to be a very high speed.

The Land Question in Context

The main reason for this extended discussion of context is that, on the intellectual and political agenda of post-war Britain, land has been treated *out* of context. Three post-war Labour governments have legislated on landownership and values (in 1947, 1967 and 1975) as part of programmes essentially designed to manage a capitalist economy more efficiently and to achieve some redistribution. Each of these measures has been largely repealed by the subsequent government, against negligible protest, and the view is now widely held that land questions are no longer on the political agenda. It is the contention of this paper that we can now begin to see why this has been so, and thus how land questions can be given their proper place.

The interpretations which follow draw upon a diverse range of work – the experience of political campaigns, well-grounded empirical work, historical analyses and some purely theoretical work: in fact upon the kinds of endeavour reported elsewhere in this book. From them we can begin to piece together the elements which would be essential to a consistent Marxist formulation of the land question.

Land, Production and Distribution

All the recent British legislation on land-value questions has been concerned with the distributional question – who should get the profits from land development? Drawing on neoclassical and neo-Ricardian theory, it has been held that landowners derive their rents or development profits essentially at the expense of capitalists. Production and the use of the built environment carry on as they would in the absence of private landowners but the surpluses are differently distributed.

A Marxist analysis shows that this is a fundamentally wrong perspective. The relations of private landownership, in their successively changing forms, do actually play a part in determining what production processes do take place, as Fine demonstrates for the coal industry (in this volume) and as Ball has shown for the speculative house-building process (1983: 167-77). Similar conclusions emerge from historical analysis of the development of urban areas (see for example Clarke and Janssen 1983) and bear out the view that urban development processes can be approached with some powerful equipment (see for example Harvey 1982; Ball 1977). Landownership relations thus take their place in a system of relations between a whole range of social agents: the owners of sites, developers and builders, designers and planners, construction workers on site and the industries called upon to contribute the components of buildings and the equipments which connect them. These relations are by no means fully understood but it is clear that they can work both ways. For example Ball's work shows how important the land problem is in the struggle between builders and workers over job security. Likewise the analysis of purely 'technical' decisions about how buildings are erected has to incorporate the study of how sites come to the builder and what role the land plays in the builder's or developer's calculations (Burchell and Hill 1980; Cullen 1982; Sebestyen 1982).

The old idea that private landownership is in simple antagonism to capital cannot thus be replaced by anything equally simple. What is clear is that rent and the wider relations of ownership play far more roles than simply distributing the pre-determined surplus profits from development. (See Perrie (1984) and Edwards and others (1984) for discussions of how abstract concepts like land emerge in social struggle.)

Barriers to Accumulation

Evidently no amount of money-capital and no amount of available labour will enable capitalists to become farmers or developers unless somehow they can get access to land on which to farm or build. This is the core of the distinctive barrier which privately-owned land poses to capital. In most other spheres the accumulation process depends on the free movement of capital; here it is liable to be impeded as we see from the analysis of mining in Chapter 6 and the historical examination of legal change in Chapter 7. And once on the land, with fixed capital being formed there and variable capital being advanced for production, the capitalist is subject to the recurrent struggle to keep the profits on 'his' capital and stop the landlord progressively milking it away in rent reviews. This struggle, which played such an important role in the history of British agriculture (Massey and Catalano 1978: chapters 2 and 7) remains ubiquitous. Whatmore (1983) has shown for example how one of the attractions of agricultural land assets for pension funds is that, in addition to the rents they charge, they can assure themselves of secure returns on the money-capital they advance to tenants for improvements. (The revived analysis of agriculture is, incidentally, proving to be an indispensable complement to urban analysis, redressing the balance after some decades of exclusively urban research. (See for example Winter (1984) and Marsden (1984).)

In urban development the same relentless struggle continues – apparent in the attention paid by developers, investors and commercial tenants to the length of leases and the frequency of rent-reviews. In this perspective there is, of course, no room for the neoclassical distinction between the market for land rights and the market in rights over buildings. The idea that the combined revenue from the disposal of a building may (using some abstract principle) be split down into its elements (return on the land, return on the building works) is a fantasy. Every developer knows better – though the distinction may be used in the ritual chants involved in bargaining and valuation.

The British legislative attempts to 'nationalise' or tax development profits have never been based on this sort of dynamic view of the proceeds from landownership. Development profit has always been seen as essentially the one-off leap in the exchange value of sites when they move from one permitted use to another or where permission is granted for an intensification of use. No attempt has been made to penetrate the continuing struggle and it is suggested that the partial nationalisation or taxation of development profits (especially the most recent and weakest version, that of 1975) thus posed little real threat to the long-run interests of modern landed property.

A further and specific result can arise where the barrier to capital flows has the effect of keeping the level of surplus profit high throughout a sector of production – the circumstance in which what Marx termed absolute rents may arise. Fine's account of coal royalties is the clearest attempt yet to show this process in action and we still lack any comparable analysis for modern urban development. Any such analysis will have to take account of the activities of the state through the planning system as an integral feature of the process, particularly as it regulates scarcities in particular local and functional sub-markets. A study in this framework of the retail industry would, for example, be of great interest; here the planning system has been almost universally restrictive in the UK: high rents and presumably a high rate of appropriation of surplus correspond with low wage levels and the exploitation of a partially-marginalised workforce. Simple explanations are not to be expected and it would follow that no simple legislation would be able to resolve the conflicts in favour of any single set of social agents.

Property Assets, Buildings and Places

Within the changing fundamental context referred to earlier in this paper, land and property-related assets play an important role. Perhaps a third of the households in the country have debts secured against the titles to their houses. These debts are to building societies and banks for whose investors they constitute assets. Likewise in the non-residential sphere large sums stand in the balance sheets of investment institutions as the exchange value of the property assets they own. Even expert predictions of oversupply seem unable to stem the flow of new investment funds in to office construction in London and speculative development continues in many parts of the country even in the crisis (JLW 1980; 1982).

This phenomenon has two important consequences. The first is economic and political: property assets are an important part of the asset structure upon which the financial system is secured. The continued valorisation of these assets is thus essential to the stability of the financial system and any major socialist programme would have to reckon with this problem if it attempted to make non-trivial changes to landownership relations. The politics of such changes are complicated by the wide diffusion of beneficial interest in the system, principally through the ubiquitous pension system and through owner-occupation of housing.

The second consequence is upon the form and the location of the buildings erected to serve as sources of rent. Since the titles to these investments have to serve as tradeable commodities, it has been argued, the tendency has been for the narrowing down of the range of building forms and of locations (Hayes 1979; Yabsley 1979: 77–80). The most attractive buildings for investors are those which would appeal to a wide range of hypothetical tenants in locations with similarly wide appeal. Marginal places and bespoke buildings are unappealing. We can thus see a tendency for buildings to approach the form of pure commodities, homogeneous within clearly understood categories. Taken to its logical extreme, this would mean an investor could telephone from New York and buy 'prime London offices' with as little need for further information as when buying gold or government stock. In a number of development submarkets we can thus see the production of what we have termed elsewhere an increasingly 'abstract' space (Edwards and others 1984).

State Land Development

To a significant extent, and under a wider range of legislative powers, the state has been involved in direct landownership and development in post-war Britain, as elsewhere in western Europe. The intellectual and political basis for this activity has had much in common with the basis for land and planning legislation, and related shortcomings are to be expected. The failure to link state appropriation of land with a democratisation of land use and production decisions – or with the wider social relations of production and consumption – is one facet of the problem. Workers engaged in on-site production of public-sector developments are (with the exception of a small minority directly employed (Direct Labour Collective 1978)) subject to relations with their employers almost identical to those ruling in private development. Even the direct beneficiaries of the redistributive aims of public housing developments have to submit to what are commonly bureaucratic, patriarchal and oppressive management arrangements (Watson 1983; Merrett 1979: chapter 8; Legg and others 1980) and now to increasingly individual, market-type relations with their public landlords.

The achievements of public sector development are not negligible, and some systematic evaluation is sorely needed within the kind of framework outlined here and developed more fully elsewhere (Edwards 1984). But it is to be expected that such work would show how little is changed when the material forces determining what is produced (and how) remain essentially the forces of capitalism. It would probably confirm the view expressed by Massey and Catalano (1978: 186-90) that public control of land has been of little real meaning because of the failure to transform the politics of land use and development.

Land Use Planning and the Land Market

From time to time demands have been made for the strengthening or weakening of the land-use planning system. It will be clear from the discussion in this paper that these demands may come from diverse quarters and that their interpretation will depend very much on the conditions prevailing at the time. The effects of the planning system may at the same time be beneficial to certain elements of capital (e.g. by restricting the supply of buildings which would compete with those built and owned) and restrictive for others (e.g. by preventing the realisation of the development potential of sites owned). More broadly the predictability created by the system as a whole will tend to increase the attraction of property assets in general, while individual owners may often have an interest in gaining exceptions in respect of their own sites and breaking the rules which yield the predictability. Capital, by and large, has been good at making the planning system its own. Conservation areas and green belts are initially restrictive but they create privileged spheres for individual and corporate investment in housing and commerce. The steady resistance to out-of-town shopping centres under the British planning system may have held back technical change in retailing but it has been the counterpart to many intensive investments in old town centres effectively protected from competition. The relaxations of planning controls which are now taking place gain much of their appeal to investors from the very fact that planning has been active (and restrictive) for so long: the opportunities created by decades of infrastructure provision, decades of conservation of buildings and landscapes and decades of restriction are now to be opened up to a new round of private realisation of development surpluses.

Conclusions

It is to be hoped that no campaign will again be mounted for reforms which simply socialise into state hands the flow of rents from an essentially private land and development market, leaving market processes to distribute types of activity and users between locations, to determine land prices and the composition of the development which takes place – albeit within a framework of land-use planning. Such strategies fail essentially because they are based on a concept of landownership and rent as playing a purely distributional role in society. A more adequate view of the land problem shows how private landownership and the land market can determine what gets built and the character of the production processes involved as well as the distribution of the proceeds.

The following suggestions are offered as the minimum characteristics which the next legislative attack on the land question should have.

(1) Since changed landownership relations do not of themselves change the essentially capitalist character of production relations, changes in both spheres must go together. It is clear that state development does not necessarily release construction workers from hazards, from job-insecurity or from the more subtle forms of subordination to capital which condition their work.

(2) As many tenants of public housing and many users of public services have found, public ownership does not of itself lead them towards individual or collective control, as users, over either production or delivery of services. If social control of land use were to have real meaning it would, in part, come through the development of new criteria and concepts of social need which could stand as the working alternative to market allocation criteria. Neither the researchers nor the professionals of the planning system have done much to develop such concepts (though see GLC 1982). While it may be said that the principles and practice of a democratised planning would emerge only in practice, this is usually a ritual excuse for inactivity.

(3) The future incomes of today's workers – after retirement – would need to be linked explicitly and directly to the labour of future generations, removing the indirect link via the long-run valorisation of capital and landed assets. If major changes were to occur in the operation of the land market it would follow that the role of land assets in the financial system would have to be dealt with at the same time.

Acknowledgement

This chapter draws heavily on the ideas and achievements of others and on discussions with many people, particularly in the Bartlett International Summer Schools. The assembly is entirely the author's responsibility.

References

Ball, M. (1977) 'Differential Rent and the Role of Landed Property' in *International Journal of Urban and Regional Research, 1*, 380–403

—— (1983) *Housing Policy and Economic Power: the political economy of owner occupation*, Methuen, London

Bentivegna, V. and M. Edwards (1983) 'The Spatial Distribution of Accumulation in Construction' in *BISS*, 4.2–4.7

BISS (1980–) *The Production of the Built Environment, Proceedings of the Bartlett International Summer School*, Bartlett School, University College London. (Dates given are for publication, usually the year following the meeting.) The 1984 volume is published by the Faculty of Architecture, University of Geneva

Budd, A. (1978) *The Politics of Economic Planning*, Fontana/Collins, London

Burchell, S. and A. Hill (1980) 'Building Contractors: forms of calculation and the contracting process' in Political Economy of Housing Workshop, *Housing, Construction and the State*, Conference of Socialist Economists, London, pp. 15–20

Clarke, L. and J. Janssen (1983) 'On a theoretical approach to the study of labour in building and construction' in *BISS*, 7–38

Coakley, J. and L. Harris (1983) *The City of Capital: London's Role as a Financial Centre*, Blackwell, Oxford

Cullen, A. (1982) 'Speculative Housebuilding in Britain: some notes on the switch to timber-frame production methods' in *BISS*, 4.12–4.18

Currie, D. and R. Smith (1981) 'Economic Trends and Crisis in the UK Economy' in Currie and Smith (eds.), *Socialist Economic Review*, Merlin, London

Direct Labour Collective (1978) *Building with Direct Labour: Local Authority Building and the Crisis in the Construction Industry*, Conference of Socialist Economists, London

—— (1980) *Direct Labour under Attack*, The Collective, London

Edwards, M. (1984) 'New Towns: Reflections on Financing and Design' in *BISS* (in press)

—— and others (1984) 'Report to plenary meeting from the working group on land and planning' in *BISS* (in press)

GLC (1982) 'A Socialist Greater London Council in Capitalist Britain?' in *Capital and Class, 17*

Glyn, A. and J. Harrison (1980) *The British Economic Disaster*, Pluto, London

Harris, N. (1983) *Of Bread and Guns*, Penguin, Harmondsworth

Harvey, D. (1982) *The Limits to Capital*, Blackwell, Oxford

Hayes, W.J. (1979) *Pension Fund and Life Assurance Company Property Investment: Origins and Effects*, unpublished MPhil thesis, University of London

Hodgkinson, S.F. (1982) *Housing Associations and the State*, unpublished MPhil thesis, University of London

JLW: Jones Lang Wootton (1980) *Offices in the City of London: a Special Report*, London

—— (1982) *Central London Offices Research: Monitoring Report – December*, London

Joint Trades Councils (1982) *State Intervention in Industry: a Workers' Inquiry*, London

Legg, C., A. Kay, J. Mason and K. Nicholas (1980) *Could Local Authorities be better Landlords?* (and case studies in Manchester, Coventry, Brent and Guildford) City University, London

Marsden, T. (1984) 'Landownership and Farm Organisation in Capitalist Agriculture' in T. Bradley and P. Lowe (eds.), *Locality and Rurality: Economy and Society in Rural Regions*, Geo-Books, London (in press)

Massey, D. (1979) 'In what sense a regional problem?' in *Regional Studies, 13*, 233-43

—— (1983) 'Industrial Restructuring as Class Restructuring: production decentralisation and local uniqueness' in *Regional Studies, 17*, 2, 73-89

—— and A. Catalano (1978) *Capital and Land: Landownership by Capital in Great Britain*, Arnold, London

Merrett, S. (1979) *State Housing in Britain*, Routledge, London (Chapter 8 by Fred Gray)

Millwood, T. (1981) 'State Expenditure in the 1970s' in D. Currie and R. Smith (eds.), *Socialist Economic Review*, Merlin, London

Minns, R. (1980) *Pension Funds and British Capitalism*, Heinemann, London

—— (1983) 'Pension Funds: an alternative view' in *Capital and Class, 20*, 104-15

Murray, R. (1983) 'Pension Funds and Local Authority Investments' in *Capital and Class, 20*, 89-102

Palmer, J.A.D. (1972) Introduction to the British edition of R. Goodman *After the Planners*, Penguin, Harmondsworth

Perrie, W. (1984) 'Time and the Subject' in *BISS* (in press)

Sebestyen, G. (1982) 'Industrialisation of Housing' in *BISS*, 4.1-4.3

Sutcliffe, B. (1983) *Hard Times: the World Economy in Turmoil*, Pluto, London

Watson, S. and H. Austerberry (1983) *Women on the Margins: a Study of Single Women's Housing Problems*, City University, London

Whatmore, S. (1983) *Financial Institutions and the Ownership of Agricultural Land*, unpublished MPhil thesis, University of London

Whitfield, D. (1983) *Making it Public: Evidence and Action against Privatisation*, Pluto, London

Winter, M. (1984) 'Agrarian Class Structure and Family Farming' in T. Bradley and P. Lowe (eds.), *Locality and Rurality: Economy and Society in Rural Regions*, Geo-Books, London (in press)

Yabsley, S.J. (1979) *Institutional Investment in Urban Development*, unpublished MPhil thesis, University of London

SELECT BIBLIOGRAPHY

There is a voluminous literature on landed property, land rent and Marxist theory. This bibliography aims to provide a basic introduction to it, concentrating on the use of rent theory in understanding the contemporary role of landed property. Works are grouped into broad subject areas. As most chapters in this book have been concerned with landed property in urban areas, greater emphasis is placed on that subject area. We have attempted to draw on the literature from English-speaking countries, especially Britain and the USA, and from France and Italy. Publication dates are given for recent editions, rather than for first editions, in the case of early works.

Major Writings on Rent Prior to 1914

Engels, F. (1975a) *The Housing Question*, Progress Publishers, Moscow

—— (1975b) *The Condition of the Working Class in England in 1844*, reprinted in *Marx–Engels Collected Works, Vol. 4*, Lawrence and Wishart, London

Kautsky, K. (1970 ed.) *La Question Agraire*, Maspero, Paris (English translation not available, but see J. Banaji (1976) 'Summary of Selected Parts of Kautsky's "The Agrarian Question" ', *Economy and Society, 5*, 1-49)

Marx, K. (1974) *Capital, III*, Preface and Parts VI and VII, Lawrence and Wishart, London

—— (1968) *Theories of Surplus Value, II*, Progress Publishers, Moscow

Land Rent and Theories of Agriculture and Underdevelopment

Allione, M. (1970) 'Metamorfosi della Rendita', *Il Manifesto, 3-4*, 41–5

Amin, S. (1977) 'Capitalism and Ground Rent' in S. Amin, *Imperialism and Unequal Development*, Monthly Review Press, New York

—— (1974) 'Le Capitalisme et la Rente Foncière: la Domination du Capitalisme sur l'Agriculture' in S. Amin and K. Vergopoulos, *La Question Paysanne et le Capitalisme*, Edition Anthropos IDEP, Paris

Emmanuel, A. (1972) *Unequal Exchange: A Study of the Imperialism of Trade*, New Left Books, London

de Janvry, A. (1981) *The Agrarian Question and Reformism in Latin America*, Johns Hopkins Press, London

Murray, R. (1977) 'Value and Theory of Rent: Part One', *Capital and Class, 3*, 100–22

—— (1978) 'Value and Theory of Rent: Part Two', *Capital and Class, 4*, 11-33

Postel-Vinay, G. (1974) *La rente foncière dans le capitalisme agricole. Analyse de la voie 'classique' du développement du capitalisme dans l'agriculture à partir de l'example du Soissonnais*, Maspero, Paris

Ray, P. (1973) *Les Alliances de Classes*, Maspero, Paris

Tribe, K. and Hussain, A. (1980) *Marxism and the Agrarian Question, Vol. I* 'German Social Democracy and the Peasantry, 1890-1907', *Vol. II* 'Russian Marxism and the Peasantry, 1861-1930', Macmillan, London

Vergopoulos, K. (1974) 'Capitalisme Difformé: le cas de l'Agriculture dans le

Capitalisme' in S. Amin and K. Vergopoulos, *La Question Paysanne et le Capitalisme*, Edition Anthropos IDEP, Paris

Rent and Landed Property: the Contemporary Theoretical Debate

Ball, M. (1977) 'Differential Rent and the Role of Landed Property', *International Journal of Urban and Regional Research, 1*, 380–403
—— (1980) 'On Marx's Theory of Agricultural Rent: a Reply to Ben Fine', *Economy and Society, 9*, 304–39
Cutler, A. (1975) 'The Concept of Ground-Rent and Capitalism in Agriculture', *Critique of Anthropology, 4/5*, 72–89
—— and Taylor, J. (1972) 'Theoretical Remarks on the Theory of the Transition from Feudalism to Capitalism', *Theoretical Practice, 6*, 20–31
Edel, M. (1974) *The Theory of Rent in Radical Economics*, Boston Studies in Urban Political Economy Working Paper 12
Fine, B. (1979) 'On Marx's Theory of Agricultural Rent', *Economy and Society, 8*, 241–78
—— (1980) 'On Marx's Theory of Agricultural Rent: a Rejoinder', *Economy and Society, 9*, 325–31
—— (1982a) *Theories of the Capitalist Economy*, Edward Arnold, London
—— (1982b) 'Landed Property and the distinction between Royalties and Rent', *Land Economics, 58*, 338–50
Harvey, D. (1982) *The Limits to Capital*, Basil Blackwell, Oxford
Massey, D. and A. Catalano (1978) *Capital and Land: Landownership by Capital in Great Britain*, Edward Arnold, London
Nardocci, A. (1973) 'La rendita fondiaria agricola in Ricardo e Marx', *Note Economiche, 1*, 51–85
Perceval, L. (1972) 'L'analyse scientifique contemporaine de la rente foncière', *Economie et Politique, 210*, 111–24
Reboul, C. (1977) 'Déterminants sociaux de la fertilité des sols', *Actes de la Recherche en Sciences Sociales, 17-18*, 85–112
Tribe, K. (1977) 'Economic Property and the Theorization of Ground Rent', *Economy and Society, 6*, 69–88
—— (1978) *Land, Labour and Economic Discourse*, Routledge and Kegan Paul, London

Urban Rent

Alquier, F. (1971) 'Contribution à l'étude de la rente foncière sur les terrains urbains', *Espaces et Societés, 2*, 75–87
Ascher, F. et J. Lacoste (1974) *Les Producteurs du Cadre Bâti: Volume I, Les Obstables au Dévelopment de la Grande Production Industrielle dans le BTP*, Université des Sciences Socialies, UER Urbanization-Aménagement, Grenoble
—— et C. Lucas (1974) 'L'Industrie du Bâtiment: des Forces Productives à Libérer', *Economie et Politique, 236*, 58–75
Ball, M. (1983) *Housing Policy and Economic Power: the Political Economy of Owner Occupation*, Methuen, London
—— (1984, forthcoming) 'The Urban Rent Question' in M. Ball, *Building Blocks: the Social Creation of the Built Environment*, London
Bentivegna, V. (1980) 'La questione della rendita urbana nella teoria marxista contemporanea', *Critica Marxista, XVII*, 4, 145–71
Bruegel, I. (1975) 'The Marxist Theory of Rent and the Contemporary City: a

Critique of Harvey' in *Political Economy and the Housing Question*, Political Economy of Housing Workshop, Conference of Socialist Economists, London

Byrne, D. and Beirne, P. (1975) 'Toward a Political Economy of Housing Rent' in *Political Economy and the Housing Question*, Political Economy of Housing Workshop, Conference of Socialist Economists, London

Cacciari, P. (1971) 'Sulla teoria della rendita urbana', *Contropiano, 3*, 635–55

—— e S. Potenza (1973) *Il ciclo edilizio in Italia. Riforma della casa e sviluppo capitalistico in Italia negli anni Sessanta*, Officina, Roma

Carassus, J. (1978) *Quelques données et hypothèses sur la formation du prix des logements*, mémoire de DEA, Université de Paris IX-Dauphine, Paris

—— (1983) *Logement: prix et production. Eléments sur la formation du prix du logement neuf et la production du cadre bâti en France entre 1962 et 1982*, Université de Paris IX-Dauphine, Paris

Clarke, S. and N. Ginsberg (1975) 'The Political Economy of Housing' in *Political Economy and the Housing Question*, Political Economy of Housing Workshop, Conference of Socialist Economists, London

Dichervois, M. and B. Theret (1979) *Contribution à L'étude de la Rente Foncière Urbaine*, Mouton, Paris and The Hague

Edel, M. (1976) 'Marx's Theory of Rent: Urban Applications', *Kapitalstate, 4-5*, 100–24, and also in *Housing and Class in Britain*, Political Economy of Housing Workshop, Conference of Socialist Economists, London

Edwards, M. and D. Lovatt (1980) *Understanding Urban Land Markets: a Review*, Vol. I of the Inner City in Context, Social Science Research Council, London

Folin, M. (1977) 'Recensione a A. Lipietz, La rendita fondiaria nella città', *Casabella, 428*, 61-2

—— (1978) 'La produzione (capitalistica) delle "condizioni generali, collettive della produzione sociale" ' in M. Folin (ed), *Opere pubbliche, lavori pubblici, capitale fisso sociale*, 21–62, Angeli, Milano

Granelle, J.J. (1970) *Espace urbaine et prix du sol*, Sirey, Paris

—— (1970) *Evolution du prix des terrains a bâtir de 1965 à 1975 dans quelques agglomérations françaises*, Fédération Nationale des Promoteurs-Constructeurs, Paris

Harvey, D. (1973) *Social Justice and the City*, Edward Arnold, London

—— (1974) 'Class-monopoly Rent, Finance Capital and the Urban Revolution', *Regional Studies, 8*, 239–55

—— and L. Chatterjee (1974) 'Absolute Rent and the Restructuring of Space by Government and Financial Institutions', *Antipode, 6*, 1, 22-36

Ive, G. (1974) 'Walker and the "New Conceptual Framework" of Urban Rent', *Antipode, 7, 1*, 20–30

Lamarche, F. (1972) 'Les fondements économiques de la question urbaine', *Sociologie et Sociétés, 4*, 15–41

—— (1976) 'Property Development and the Economic Foundations of the Urban Question' in C. Pickvance (ed.), *Urban Sociology: Critical Essays*, Tavistock Publications

Lipietz, A. (1974) *Le tribut foncier urbain. Circulation du capital et propriété foncière dans la production du cadre bâti*, Maspero, Paris

Lisanti, G. (1973) 'Città e rendita fondiaria', *Problemi del Socialismo, 15*, 368–93

Lojkine, J. (1971) 'Y a-t-il une Rente Foncière Urbaine?', *Espaces et Societés, 2*, 89–94

—— (1977) *Le Marxisme, l'Etat et la question urbaine*, Presses Universitaires de France, Paris

Markusen, A. (1978) 'Class, Rent and Sectoral Conflict: Uneven Development in Western US Boomtowns', *Review of Radical Political Economy, 10, 3*, 117–29

Nicolini, R. (1972) 'Rendita e creazione del plusvalore nel settore edile in Italia dalla riconstruzione ad oggi', *Controspazio, 8–9*, 50–7

Scott, A. (1976) 'Land and Land Rent: an Interpretative Review of the French Literature', *Progess in Geography*, *9*, 103–45

—— and S. Roweis (1977) 'Urban Planning in Theory and Practice: a Reappraisal', *Environment and Planning A*, *10*, 229–31

—— (1980) *The Urban Land Nexus and the State*, Pion, London

Topalov, C. (1970) *Capital du propriété foncière. Introduction à l'étude des politiques foncières urbaines*, Centre de Sociologie Urbaine, Paris

—— (1974) *Les promoteurs immobiliers. Contribution à l'analyse de la production capitaliste du logement en France*, Mouton, Paris

—— (1977) 'Surprofits et rentes foncières dans la ville capitaliste', *International Journal of Urban and Regional Research*, *1*, 425–46

—— (1984) *Le profit, la rente et la ville: éléments de théorie*, Economica, Paris

Walker, R. (1974) 'Urban Ground Rent: Building a New Conceptual Framework', *Antipode*, *6*, 1, 51–8

—— (1975) 'Contentious Issues in Marxian Value and Rent Theory: a Second and Longer Look', *Antipode*, *8*, 31-53

NOTES ON CONTRIBUTORS

Michael Ball is a lecturer in economics at Birkbeck College, University of London. He has written widely on housing, the construction industry and land market questions.

Vincenzo Bentivegna is a professor in the Instituto di Urbanistica of the Faculty of Architecture, University of Florence. His research has ranged widely over issues in urban economics and particularly economic methods used in planning. He is currently working on a textbook on rent and has been centrally concerned with the Bartlett International Summer Schools on the Production of the Built Environment.

Michael Edwards is a lecturer in the economics of planning at the Bartlett School of Architecture and Planning, University College London. His work and teaching have mainly focused on land-related aspects of planning, but with a subsidiary interest in local labour-market analysis. In the framework of the Bartlett International Summer Schools, he is currently working with others on the preparation of educational material on the Production of the Built Environment.

Ben Fine is a reader in economics at Birkbeck College, University of London. He has written extensively on Marxist economic theory and undertaken detailed research on the development of the British coal industry.

Marino Folin is Professor in the Department of Social and Economic Analysis of the Environment, Faculty of Architecture, University of Venice. He has worked on the form and nature of town planning and is now doing research on housing and public works expenditure in European countries. His books include *La città de capitale* (1972, 1976, translated into Spanish and Dutch) and *Opere pubbliche, lavori pubbliche, capitale fisso soziale* (1979).

Ambrois Gravejat is a professor at the Centre de Recherches Economique, University of Saint-Etienne. His major published work is *La Rente, le Profit et la Ville* (1980). In the framework of the Bartlett International Summer School on the Production of the Built Environment he is collaborating on European comparative studies of the construction sector.

Alain Lipietz is a senior researcher at the Centre des Etudes Prospectives d'Economie Mathematique Appliquée à la Planification (CEPREMAP) in Paris. His publications include *Le Tribut Foncier Urbain* (1974) and *Le Capital et son Espace* (1979). He is now working on the international division of labour including the steel and car industries.

Michael McMahon is a Canadian researcher at the Bartlett School, University College London. He completed his MPhil thesis on *Town Planning and the*

Development of a Land Market in Britain, 1845–1910 in 1982 and is now working on his doctorate which concerns the development process and the planning system in Canada.

Agostino Nardocci is a lecturer in economics in the Department of Social and Economic Analysis of the Environment of the Faculty of Architecture, University of Venice. He is writing a book on rent in classical economics and in Marx's work and has previously published a number of papers on the theory of rent.

Christian Topalov is a senior researcher at the Centre de Sociologie Urbaine in Paris. He has published *Capital et Propriété Foncière* (1973) and *Les Promoteurs Immobilières* (1974) and is now working on comparative social policies in the 1920s and 1930s in European countries and the USA.

INDEX

References to Tables and Figures are identified by (T) and (F) respectively following the page number.